Impedimetric Biosensors for Medical Applications

Current Progress and Challenges

Jo V. Rushworth

Natalie A. Hirst

Jack A. Goode

Douglas J. Pike

Asif Ahmed

Paul A. Millner

ASME Press books are available at special quantity discounts to use as premiums or for use in corporate training programs. For more information, contact Special Sales at infocentral@asme.org

A catalog record is available from the Library of Congress.

Print ISBN: 978-0-7918-6024-3
ASME Order No. 860243
Electronic ISBN: 978-1-60650-637-0

Series Editors' Preface

Biomedical and Nanomedical Technologies (B&NT)
This concise monograph series focuses on the implementation of various engineering principles in the conception, design, development, analysis and operation of biomedical, biotechnological and nanotechnology systems and applications. The primary objective of the series is to compile the latest research topics in biomedical and nanomedical technologies, specifically devices and materials.

Each volume comprises a collection of invited manuscripts, written in an accessible manner and of a concise and manageable length. These timely collections will provide an invaluable resource for initial enquiries about technologies, encapsulating the latest developments and applications with reference sources for further detailed information. The content and format have been specifically designed to stimulate further advances and applications of these technologies by reaching out to the non-specialist across a broad audience.

Contributions to *Biomedical and Nanomedical Technologies* will inspire interest in further research and development using these technologies and encourage other potential applications. This will foster the advancement of biomedical and nanomedical applications, ultimately improving healthcare delivery.

Editor:
Ahmed Al-Jumaily, PhD, Professor of Biomechanical Engineering & Director of the Institute of Biomedical Technologies, Auckland University of Technology.

Associate Editors:
Christopher H.M. Jenkins, Ph, PE, Professor and Head, Mechanical & Industrial Engineering Department, Montana State University.

Guy M. Genin, PhD, Associate Professor of Mechanical Engineering and Materials Science, Washington University in St. Louis, and Associate Professor of Neurological Surgery, Washington University School of Medicine.

Feng Xu, PhD, Professor, The Key Laboratory of Biomedical Information Engineering of Ministry of Education, School of Life Science and Technology, and Director, XJTU Biomedical Engineering & Biomechanics Center, Xi'an Jiaotong University, China.

Contents

Abstract

In this monograph, the authors discuss the current progress in the medical application of impedimetric biosensors, along with the key challenges in the field. First, a general overview of biosensor development, structure and function is presented, followed by a detailed discussion of impedimetric biosensors and the principles of electrochemical impedance spectroscopy. Next, the current state-of-the art in terms of the science and technology underpinning impedance-based biosensors is reviewed in detail. The layer-by-layer construction of impedimetric sensors is described, including the design of electrodes, their nano-modification, transducer surface functionalisation and the attachment of different bioreceptors.

The current challenges of translating lab-based biosensor platforms into commercially-available devices that function with real patient samples at the POC are presented; this includes a consideration of systems integration, microfluidics and biosensor regeneration. The final section of this monograph describes case studies of successful impedance-based biosensors for the detection of a range of analytes from small molecules up to whole microorganisms. Finally, the authors put forward future perspectives for the clinical applications of impedimetric biosensors.

1.   Introduction and scope

A biosensor is a compact analytical device, often considered as a "lab-on-a-chip", which facilitates the detection and quantification of a target analyte, at the point-of-care (POC). This confers several key advantages over laboratory-based means of analyte detection: biosensors are rapid, cost-effective and operate at the bedside or in the home without the need for specialist users or equipment. As such, biosensors can save lives and tackle medical problems sooner by providing a faster diagnosis. Biosensors prove particularly useful, in addition to their application in hospitals and GP surgeries, (a) in the developing world, where cheap disease detection without the need for specialist clinicians is lacking and, (b) in the developed world, for self-monitoring of diabetes, cholesterol levels etc.

The first and best known biosensor is the blood glucose monitor, which is used by patients with diabetes to measure blood sugar levels. Since the glucose sensor was first demonstrated 50 years ago by Clark and Lyons (Clark and Lyons, 1962), the world market for self-monitoring of blood glucose (SMBG) using biosensors has grown exponentially to $8.8 billion per year in 2008 (Hughes, 2009). The World Health Organisation forecasts that one in ten adults will have diabetes by 2030 and, therefore, the market for glucose biosensors is set to continue to grow. The requirement for POC biosensors to detect a wide range of targets is now increasing; in a medical context, biosensors have been developed which can pick up minute quantities (as low as femtomols) of various ions, metabolites, proteins, enzymes, steroids, nucleic acids and microorganisms (Vo-Dinh and Cullum, 2000). These substances are often markers of a specific disease state, for instance cardiac biomarkers to diagnose a heart attack (Billah et al., 2008) or viruses to indicate pathogenic infection (Caygill et al., 2010). Biosensors are also useful and effective tools in other public health fields, where *in situ* analyte monitoring is also required, for instance in environmental water monitoring for radiation (Conroy et al., 2010) and pesticides (Vakurov et al., 2005) and the detection of antibiotics and other substances in food products (Tsekenis et al., 2008).

Typically, biosensors for medical applications are designed to detect the substance of interest in a patient sample, such as urine, blood, saliva or faeces. Additionally, biosensor devices for the detection of volatile substances in the air (i.e. in a patient's breath), known as electronic noses or e-noses, have been employed for biomedical applications, such as the detection of cancerous cells (Roine et al., 2012), as well as in pharmaceuticals, food and beverages and agriculture (Baldwin et al., 2011; Wilson, 2013; Wilson and Baietto, 2011). Implantable biosensors are currently being developed that can be placed within the patient's body to allow the continuous, real-time monitoring of analytes such as glucose (Wilson and Gifford, 2005). Whilst these systems provide better data and negate painful finger-prick or blood tests, they present new hurdles, such as minimising biofouling of the implanted device, the challenge of powering and receiving data from the system and preventing immune responses against the foreign-body sensor system (Wang et al., 2013b).

Although biosensor systems are highly diverse, the basic principles underpinning biosensor design and function are universal. A biosensor is an analytical tool which couples a biorecognition event - analyte binding to a specific bioreceptor – to a physico-chemical transducer system to generate a measurable output signal, the strength of which is proportional to the level of analyte. Biosensors can be categorised according to the type of biorecognition element (e.g. enzymes, antibodies, nucleic acids etc.) or by the method of signal transduction (e.g. electrochemical, optical or mechanical).

A notable advance in the field of medical biosensing has been the advent of novel signal transduction methods and bioreceptors in order to facilitate the detection of a wider range of analytes, with greater sensitivity. The biorecognition element of the first biosensors, such as the blood glucose sensor, relied upon enzyme activity to alter levels of a redox-active product at the electrode. This limited the range of analytes to those which were subject to enzymatic action. Nowadays, we are no longer limited to the detection of redox-active or charged species. Electrochemical impedance spectroscopy (EIS) is an electrochemical technique which exploits the change in charge-transfer resistance or capacitance that occurs at the biosensor surface upon analyte binding to the biorecognition element (Caygill et al., 2012). Thus, impedimetric biosensors negate the need for a redox-active or charged analyte, so a wide range of bioreceptors – including antibodies, aptamers, nucleic acids and even whole cells – can now be employed to detect practically any analyte. Impedimetric biosensors also offer key advantages such as low cost, high sensitivity and ease of miniaturisation. As such, impedimetric biosensors present a very favourable method of POC diagnostics for a wide range of medically-relevant analytes.

Impedimetric biosensors can offer the high sensitivity, low cost and ease of miniaturisation required to detect very low concentrations of clinically-relevant analytes in patient samples (Millner et al., 2012). These analytes include small molecules (Chullasat et al., 2011), protein biomarkers (Billah et al., 2010), nucleic acids (Yang et al., 2013b) and whole pathogenic organisms such as viruses (Caygill et al., 2012) and bacteria (Wang et al., 2012).

Nevertheless, although impedance spectroscopy is an established technique that was developed in the late 19th century (Heaviside, 1894a), its application toward biosensors was only realised in the 1980s (Newman et al., 1986; Taylor et al., 1988). Therefore, examples of impedimetric biosensors that have reached the clinic are currently lacking. In spite of recent progress in developing biosensors for medical applications, many biosensor platforms still remain rooted in the laboratory, where analyte monitoring in defined buffer solutions is easily achievable. In contrast, these systems often fail to function well in *bona fide* patient samples such as serum, blood, urine, sweat and faeces, which are complex matrices containing a lot of non-specific material which interferes with analyte detection. The translation of biosensing technology from lab to clinic is a key challenge in the field (Daniels and Pourmand, 2007). This requires sufficient reduction of background signals from non-specific interactions to allow analyte detection in real samples, and often requires signal amplification to allow the detection of clinically-relevant levels of analyte. Other important considerations include the cost of materials, reliability of the system, stability of the bioreceptors under field conditions and, in some cases, the capacity to re-use the sensor chips.

Another challenge is that proof-of-concept biosensor systems are typically developed in a laboratory using bulky, specialist equipment in order to optimise the system parameters and to validate the technique in artificial samples. Then, the goal is to miniaturise the system to produce a portable, hand-held device that can be commercialised in order to bring the biosensor system into the field.

In this monograph, we discuss the current progress in the medical application of impedimetric biosensors, along with the key challenges in the field. First, a general overview of biosensor development, structure and function is presented, followed by a detailed discussion of impedimetric biosensors and the principles of EIS. Next, the current state-of-the art in terms of the science and technology underpinning impedance-based biosensors is reviewed in detail. The layer-by-layer construction of impedimetric sensors is described, including the design of electrodes, their nano-modification, transducer surface functionalisation and the attachment of different bioreceptors. The current challenges of translating lab-based biosensor platforms into commercially-available devices that function with real patient samples at the POC will be presented; this includes a consideration of systems integration, microfluidics and biosensor regeneration. The final section of this monograph describes case studies of successful impedance-based biosensors for the detection of a range of analytes from small molecules up to whole microorganisms. Finally, we put forward future perspectives for the clinical applications of impedimetric biosensors.

2. Biosensors

2.1 Brief history of biosensor development

A biosensor in its most simple form may be described as a device comprising three parts: a biological recognition system, a transducer, and a signal processing display (Conroy et al., 2009). Interaction of the analyte of interest with the biorecognition element is converted to a measurable signal by the transducer, before conversion to the readout or display (Vo-Dinh and Cullum, 2000). The basic structure of a biosensor is illustrated in Figure 2-1.

The three basic components of a biosensor are a biorecognition element, a transducer and a readout display. Analyte binding to the bioreceptors within the biorecognition element causes a difference in signal to be transduced through to the readout display.

Biosensors were first developed in the 1960's, with the first biosensor described by Clark and Lyons in 1962 at the Cincinnati Children's Hospital for monitoring during cardiovascular surgery. Their original method utilised an oxygen electrode with glucose oxidase entrapped between semi-permeable dialysis membranes. The electrode measured the oxygen consumption during the enzyme reaction, to consequently give a glucose concentration (Clark and Lyons, 1962). Errors caused by variation in O_2 levels in the solution however, led to the reaction product hydrogen peroxide becoming the product of interest to be measured. These so called "first generation" of amperometric sensors developed directly measured the electroactive species enzymically produced or consumed. Second generation biosensors were developed in the 1980's to overcome the requirement for a high potential in order for measurement of the electroactive species, which decreased specificity (Wang, 2001). The use of electron mediators such as ferrocene, to shuttle electrons to the electrode from the enzyme, allowed a lower working potential to be used and thus decreased interference from other redox species present. Mediators may be free in the electrolyte, or immobilised on to the working electrode with the enzyme. The development of third generation biosensors in the 1990's progressed biosensing technology further in allowing direct, unmediated electron transfer between the enzyme's redox centre and the working electrode surface, by use of an enzyme capable of direct electron transfer such as horseradish peroxidise or cytochrome c. Clearly this may only be achieved with very small enzymes which have redox centres close to their surface, or those whose structure is appropriate.

In recent times biosensors have progressed from being predominantly enzymatic to encompass a wide variety of bioreceptors such as DNA, antibodies and aptamers, and using a range of transduction methodologies including optical, piezoelectric, and within electrochemical approaches an increased emphasis on impedimetric immunological sensing (Song et al.,

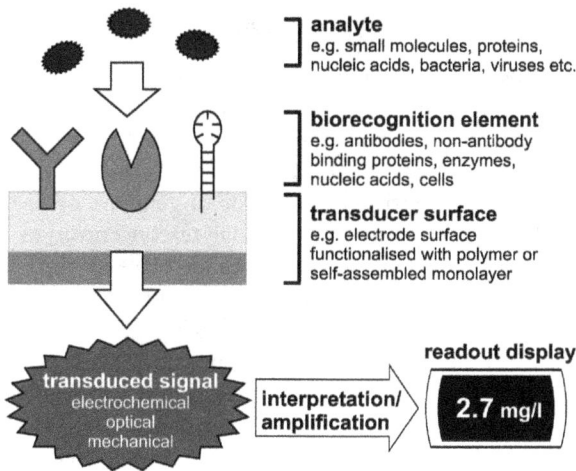

Figure 2-1 Schematic of a biosensor.

2006). Each approach can be tailored to solving the specific challenges of the requirement to be met, for example the use of impedance sensors to measure non-electroactive species, to which amperometric biosensors are confined. Generic areas of ongoing biosensing research include design of integrated systems to allow multiplexed sensing, miniaturisation, and methods of continually improving sensitivity, selectivity and stability (Turner, 1997). Work to date also has a focus on the development of electrode surfaces which not only help facilitate electron transfer but also provide a structural matrix with which to immobilise biorecognition molecules (Millner et al., 2009). These modified surfaces are discussed later (Section 4.2.1).

2.2 Applications of biosensors

Since the pioneering work of Clark and Lyons, there have been great developments in sensing technologies research, for a wide variety of applications including environmental monitoring, food and water quality control and medical diagnostics and treatment. The advantages of adaptability, point-of-care, speed, portable sampling, low cost and ease of use have rendered biosensing an important alternative to standard centralised and sophisticated bioanalytical systems (Cosnier, 2005). Applications are numerous, and areas of growth are particularly seen in the clinical setting, where DNA biosensors lend themselves to detection of genetic diseases. The detection of antigenic proteins including bacteria, viruses and parasites may be achieved at low concentration by antibody-based biosensors with high specificity. Currently, the blood glucose biosensor is still the most widespread example of a biosensor, now accounting for approximately 85% of the biosensor market globally at an estimated $8.8 billion (Hughes, 2009).

2.3 Biosensor architecture – an overview

Biosensors may be classified either by their biological recognition element e.g. antibody, enzyme, DNA, or by their type of transducer (Vo-Dinh and Cullum, 2000). Electrochemical transducers are the most commonly used type in biosensing (Conroy et al., 2009). They utilise a biological event from analyte interaction with a bioreceptor to generate or to modulate an electrical signal which is related to the analyte concentration (Ronkainen et al., 2010). This biological interaction takes place at the working electrode, where enzyme, antibody etc is immobilised to allow electron transfer directly at its surface, generating current. Working electrodes are commonly noble metals such as gold, or other conducting substances such as carbon. A two, or commonly three, electrode closed system is used, in which current flows between the working and counter electrodes, with respect to the reference electrode, and is measured. Finally, the complete electrochemical cell must contain an electrolytic medium, capable of carrying the ionic charges (Korotcenkov, 2010); in a clinical setting, this is usually the fluid in which the analytes are found e.g. blood, urine etc.

There are many different transducer surfaces used in biosensing to which the bioreceptors can be tethered. Generally, these may be divided into two categories: insulating films, and polymer matrices (Millner et al., 2009). Insulating film based surfaces include self assembled monolayers (SAMs). SAMs consist of an ordered layer of molecules with a functional head group with affinity for the surface substrate, and a hydrophobic tail group facing into the solution. SAMs may then be modified as mSAMs – mixed self assembled monolayers – by addition of functionalised lipids or cross-linkers, in order to construct a platform for immobilisation. The biotin-avidin system is one way by which this can be achieved, where tetravalent avidin is able to link the biotinylated bioreceptor to the biotin-tagged surface. This method has been previously been used in biosensors for detection of haemoglobin (Hays et al., 2006).

Conducting polymer matrix surfaces are typically formed on the working electrode surface by electropolymerisation using cyclic voltammetry (Millner et al., 2009), which allows fine control of the surface thickness. The formed polymer layer can then be used to immobilise electrochemical mediators and/or enzymes and other biological recognition elements by electrostatic absorption, covalent bonding, biotin-avidin coupling, or other methods (Gerard et al., 2002). Alternatively, proteins may be entrapped within the polymer matrix itself, although this can reduce activity due to decreased diffusion of analyte through the polymer layer to the redox site. Common polymers used are formed from monomers such as pyrrole, aniline and their derivatives, as they are easily electropolymerised at low redox potentials and have a high stability. Transducer surfaces are discussed in more detail in Section 4.2.

2.4 Biorecognition element

The biorecognition element of a biosensor consists of a bioreceptor which is attached to a matrix support on a transducer surface. The bioreceptor is chosen to specifically interact with an analyte or material of interest, which leads to transduction and signal generation (Katz and Willner, 2003). Bioreceptors are commonly proteins such as enzymes, antibodies, cellular receptor proteins, non-antibody binding proteins and antigens, or nucleic acids such as oligonucleotides and DNA or RNA aptamers. The properties of the bioreceptor are of great importance, as they essentially confer the sensitivity and specificity of the overall biosensor (Conroy et al., 2009; Hock et al., 2002). Other important considerations are the method of immobilisation of the bioreceptors onto the supportive matrix, and their correct orientation, homogeneity and stability (Cosnier, 2005).

2.4.1 Enzymes

Enzymes as bioreceptors are arguably the most common type, doubtless due to their use in the vast market of glucose monitoring (Luong et al., 2008), although measurement of other analytes such as cholesterol and lactate are emergent systems, using the relevant oxidase enzymes. They are also relatively easy to use and to attach onto transducer surfaces, conferring efficient biocatalytic activity, rendering them a popular option (Miscoria et al., 2006). A native enzyme may be used, in which the concentration of analyte is equal to the enzyme substrate, or the analyte may function as an enzyme inhibitor (Pohanka and Skladai, 2008). Affinity based sensors may also be constructed, by the use of enzymes as labels bound to antibodies, antigens and oligonucleotides with a specific sequence. The oxidoreductases are the most widely used family of enzymes for electrochemical application (Pohanka and Skladai, 2008; Ricci and Palleschi, 2005) with glucose oxidase the most common, as mentioned previously, for glucose monitoring (Wilson and Turner, 1992). Other examples of oxidase enzymes are glucose dehydrogenase, lactate oxidase and dehydrogenase for lactate detection and alcohol oxidase for ethanol. Glucose oxidase catalyses the conversion of glucose and oxygen to gluconic acid and hydrogen peroxide, of which the hydrogen peroxide product may be detected amperometrically *via* an oxidative or reductive current signal (Wang, 2008). Fundamental problems with this direct method include interference from electroactive substances, such as ascorbic acid (vitamin C), which undergo redox reactions at relatively low working potentials and thus reduce selectivity of the biosensor. The use of mediators such as ferrocene and Prussian Blue to allow low potential, selective detection of hydrogen peroxide and other reaction products has since found wide use to overcome this difficulty (Karyakin et al., 1995; Wilson and Turner, 1992). Another key limitation of enzymes as bioreceptors is the necessity for the analyte of interest to be a substrate for an enzymatic reaction, of which the products can then be measured at the transducer. This restricts the range of enzyme biosensors that can be constructed. Other electrochemical transducer systems such as immunosensors utilising antibodies as bioreceptors can be fabricated against almost any analyte, and are discussed below.

2.4.2 Antibodies

Antibodies, or immunoglobulins, (IgG, IgM, IgA, IgE) are produced as part of the physiological vertebrate immune response to pathogenic organisms and toxins (Holliger and Hudson, 2005). IgG is the main serum antibody in mammals and is the immunoglobulin that is almost exclusively used in clinical therapeutics. Antibodies have been recognised for potential use as biosensor bioreceptors since the late 1980's (Conroy et al., 2009). These "immunosensors" using antibody based biorecognition have since been developed on a multitude of transducer surfaces to measure a wide range of analytes. A main advantage is the potential for highly selective biosensing, as the antibody can undergo very specific binding to the chosen analyte of interest, with high affinity (Conroy et al., 2009). This is particularly the case with monoclonal antibodies, which are derived from only one B lymphocyte cell line, although these present much higher production costs. Polyclonal antibodies originate from multiple B lymphocyte cell lines and are more commonly used. Despite broader specificity, or multiple epitope recognition, they are more tolerant to variability in antigen structure, often display higher avidity, and are less expensive (Zourob et al., 2008). Recombinant antibodies, genetically engineered for purpose, are finding increased use in immunosensor development, as they can be modified to allow better selectivity, size, stability and easier immobilisation onto the transducer surface (Holliger and Hudson, 2005). They are set to be the next generation of key diagnostic and therapeutic biosensor tools in the clinical setting, targeting cancer, inflammatory, autoimmune and viral diseases.

2.4.3 Non-antibody binding proteins

As discussed, antibody based biosensors exploit the specificity of the immunological reaction between analyte and antibody. However in a clinical setting, there may be non-specific reactions caused by cross reactivity with non-specific antibodies if patients suffer or have been exposed to a similar disease, leading to false positive results (Soledad Belluzo et al., 2011). Enhanced results have been shown with recombinant antibodies, but also with the use of recombinant proteins. The fusion of DNA sequences encoding antigenic proteins has been used to design new epitopes for superior sensing, as well as the ability to group several of these peptides into one molecule as a highly sensitive bioreceptor. These proteins can be used to construct biosensors to detect antibodies themselves, target proteins, bacteria, viruses and parasitic organisms. They are particularly advantageous in the capacity to engineer particular functional groups onto them, such as targeted cysteine thiols, which can facilitate orientation of the bioreceptor to increase sensitivity. This concept is discussed in greater detail in Section 4.2. Non-antibody binding proteins may also reduce costs, as they are produced synthetically, with no requirement for an animal host. Bacteriophage, viruses that specifically attach to and infect bacteria, may form the biorecognition element of biosensors, with the recent advent of phage libraries facilitating their use (Meyer and Ghosh, 2010).

2.4.4 Nucleic acids

Genetic analysis is an area of increasing importance since the completion of the human genome project, and has huge implications in the diagnosis and monitoring of genetic disease as well as in detection of DNA damage and interactions. DNA biosensors are consequently gaining considerable interest as a rapid, simple, inexpensive method of gaining sequence-specific genetic information as compared to standard DNA analyses (Wang, 2000). These DNA biosensors are typically based on the use of a single stranded DNA probe immobilised onto a transducer surface, to hybridise specifically to respective base pairs (Katz and Willner, 2003). So called "gene chips" or DNA microarrays allow the multiplex analysis of numerous complex DNA samples with efficiency and precision by immobilising multiple DNA probes for analysis. The introduction of peptide nucleic acid, synthetic DNA in which the sugar phosphate backbone is replaced with pseudopeptide, provides biosensors with a high specificity up to single base mismatches, and allows greater freedom of experimental conditions (Wang, 2000). DNA dendrimer nucleic acids as biorecognition elements are also gaining interest, as their branching allows for greater hybridisation to multiple complimentary strands, giving higher signal and greater sensitivity. Advantages of nucleic acid biosensors also include relatively simple construction and the possibility of regeneration for multiple use due to the reversibility of hybridisation, thus offering lower cost (Millan and Mikkelsen, 1993).

2.4.5 Others

In principle, any molecule capable of recognising a target analyte may be exploited as a bioreceptor in biosensing. Nucleic acid aptamers are DNA or RNA sequences with three-dimensional structures first discovered in 1990, which can specifically bind to target molecules and have immense potential for biosensor medical diagnostics as well as applications including environmental monitoring (Song et al., 2008). Twenty years after discovery, aptamers have been shown to bind with high specificity to a wide range of molecules including proteins, peptides, whole cells, drugs and amino acids. They may also be fabricated readily, are small in size and cost effective with superior stability (Xiao et al., 2005). Their high affinity is derived from their ability to fold upon binding with their analyte. They are isolated *in vitro* by the Selective Evaluation of Ligands by Exponential Enrichment (SELEX) procedure, and have been dubbed "chemical antibodies". However, aptamer technology is still in development, with challenges to be overcome such as the limited availability of aptamer types and, as yet, poor knowledge of optimal immobilisation techniques onto transducer surfaces.

2.5 Transducer element

The transducer element is used to convert the biological event resulting from the interaction of the bioreceptor with analyte into a measurable signal which may then be read at the signal display (Katz and Willner, 2003). As previously mentioned, biosensors may be classified by their biological recognition element or by their type of transducer. Types of transducer include optical, electrochemical, mass-based, thermal and piezoelectric. Electrochemical transducers are the oldest and most commonly used.

2.5.1 Optical biosensors

Optical biosensors are advantageous in their immunity to electromagnetic interference, ability to sense remotely and use of multiple detection in one device (Fan et al., 2008). There are two main types: chromophore-based detection, with the use of fluorescent or absorbent tags, and label-free detection, with the target analyte detected in its natural form. Label-free detection is considered superior, as it negates the costly and time consuming tagging step which can also negatively affect molecular interactions (Cooper, 2002). Within label free optical biosensing, there are a number of detection methods including refractive index (RI) detection, optical absorption detection and Raman spectroscopy. All of these types of optical biosensors work on the same principle, that of measurement of analyte-bioreceptor interaction, in which for example the interaction changes the RI at the sensor surface which is shown optically (Fan et al., 2008). Surface plasmon resonance (SPR) is a type of RI detection optical biosensor which has been extensively investigated for DNA and protein bioreceptors. Surface plasmon resonance is a means of real time detection and is sensitive, but limited penetration confers difficulty in measurement of large molecules such as bacteria, and currently they are not suitable for multiplexed platforms as they can only detect one analyte (Homola et al., 1999).

2.5.2 Piezoelectric biosensors

Piezoelectric "mechanical" transducers act by transforming the physical mass of an analyte into an electrical signal (Janshoff et al., 2000). Primary examples include quartz crystal microbalances (QCM), surface acoustic wave devices, atomic force microscopy (AFM) and others. Quartz crystal microbalances consist of an oscillating crystal, which changes frequency in response to a change in mass at its surface due to the presence of an analyte (Muramatsu et al., 1987). These devices have traditionally been used in vacuum deposition and other industrial systems, but are gaining interest for clinical analysis fields with the use of biorecognition layers coated onto the crystal. Acoustic wave sensor devices have been used for the detection of biological and chemical entities within gas and liquid states. The sensor interface selectively absorbs molecules of interest from within the medium, which changes the amplitude, velocity and surface resonance of the device, all of which may be correlated with analyte concentration (Korotcenkov, 2010).

They have been shown to be highly sensitive, and are small and inexpensive to produce. Atomic force microscopy utilises a cantilever running over the surface of analyte, which changes the characteristics of the cantilever when in contact (Ziegler, 2004). The cantilever surface may be modified with a bioreceptor layer as with QCM, to improve selectivity. AFM is a highly sensitive modality, and has shown promise in determination of pathogen detection, DNA analysis and tumour marker detection (Lavrik et al., 2004).

2.5.3 Electrochemical biosensors

Electrochemical sensors are the largest and most developed group of biosensors, with consequently the greatest commercial success in clinical, industrial and environmental fields (Korotcenkov, 2010). General advantages are rapid response times, user friendly application, low cost, and small size with the ability to be easily miniaturised (Zelada-Guillen et al., 2013). Further classification of electrochemical transduction may be divided into amperometric, potentiometric and impedimetric biosensing.

2.5.3.1 Amperometric sensors.

Amperometric biosensors function by the direct measurement of the current produced when a constant potential is applied between two electrodes (Korotcenkov, 2010). The current itself is generated by the oxidation or reduction of electroactive species produced by the bioreceptor, commonly an enzyme, in response to the analyte. If no other redox species are present, the current generated is proportional to the target analyte concentration and therefore may be determined in samples. Oxidoreductase and dehydrogenase enzymes often generate electroactive products such as hydrogen peroxide during their catalytic cycle, and are commonly used in amperometric sensing. Glucose oxidase is the most frequently used enzyme in amperometry, and forms the biorecognition element of any medical glucose sensor (Wang, 2001). Advantages of amperometric biosensors are a rapid response time for point-of-care diagnostics and excellent sensitivity. However, they can suffer low specificity depending on the potential applied, which if high allows other redox species to contribute to the signal produced and thus give an erroneous reading (Korotcenkov, 2010). They are also limited in the range of analytes that may be measured, as these must be substrates for an enzymatic reaction.

2.5.3.2 Potentiometric sensors.

Potentiometric biosensors may almost be considered the opposite to amperometric sensors, as in this case it is the voltage produced which is measured when a constant (zero) current flows through the electrochemical cell. Ion selective electrodes are used to measure the potential, by the use of an ion-selective membrane on the electrode which defines the target ion measured (Korotcenkov, 2010). In a manner similar to amperometric sensing, the turnover of analyte by enzymes immobilised on the sensor surface gives rise to a change in concentration of a measurable ion (e.g. H^+ or NH_4^+). Potentiometry has been used for many decades for the detection of various analytes, or ions directly, and is the basis of the modern pH meter (Zelada-Guillen et al., 2013). Potentiometric biosensors confer excellent selectivity, have broad dynamic ranges and are not destructive (Gerard et al., 2002). They are inexpensive, and readily portable. Potentiometric biosensors have been shown to be useful from a pharmaceutical perspective, to measure penicillin and lysine (Parsajoo et al., 2012). However, they are slow acting, and may have low sensitivity.

2.5.3.3 *Impedimetric sensors.*

Impedance biosensors detect changes in the electrical field due to a change in capacitance and electron transfer resistance at the working electrode surface arising from analyte-bioreceptor interaction. Simply put, as the concentration of analyte increases, analyte binding to the bioreceptor increases, and subsequently impedance across the electrode surface changes, which is detected at the transducer. Impedance can increase or decrease depending upon the analyte. Impedance measurement is accomplished by the application of a small sinusoidal voltage at low frequency, followed by measurement of the resulting current and its phase shift. The current-voltage ratio gives the impedance (Daniels and Pourmand, 2007). Impedance biosensors were first described in the 1980's, and lend themselves to the measurement of a wide range of analytes, from small molecules to proteins and up to whole bacteria and viruses (Berggren et al., 2001; Katz and Willner, 2003). This is one of the key advantages of impedance sensing over other types of electrochemical transduction; as there is no requirement for electroactive species, there are virtually no limitations on analyte type. They have promising

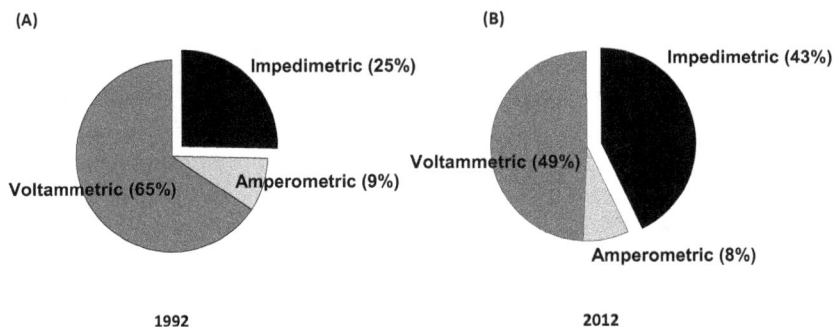

Figure 2-2 Increase in published reports of impedimetric biosensors during the last 20 years. The percentage of publications containing the key words "electrochemical biosensor" also containing either "impedance/impedimetric", "amperometry/amperometric" or "voltammetry/voltammetric" (A) in 1992 and (B) in 2012. Web of Science was used as the search engine and only "articles" were counted in the search.

importance in clinical diagnostics, pathogen detection, food quality control and environmental monitoring. Currently, no impedance biosensor has demonstrated widespread commercial success. There are still challenges with reproducibility, non-specific binding, and high limits of detection in many cases (Berggren et al., 2001; Daniels and Pourmand, 2007). However, the growing number of publications within this field shows clear prospects for resolution of these issues (Figure 2-2). Greater knowledge of immobilisation strategies and testing on real samples will improve the technology of this rapidly developing technique.

2.6 Summary

Since the first biosensors were described over 40 years ago, there has been tremendous activity towards the development of biosensing devices for applications ranging from food quality control, environmental monitoring and clinical diagnostics and therapeutics. The increased understanding of bioreceptor types such as aptamers and antibodies for sensitive and specific targeting with different immobilisation strategies to transducers ranging from electrochemical to mechanical, has led to major advances in the biosensing field. Future growth looks to continue to develop new and improved biosensors, with use of expanding technologies based on fundamental scientific principles.

3. Electrochemical impedance spectroscopy

3.1 Brief history of the development of impedimetric biosensors

The foundations of electrochemical impedance spectroscopy date back to the end of the 19th century through the work of the English mathematician and physicist, Oliver Heaviside, who coined the term "impedance" (Heaviside, 1894a, b; Macdonald, 2006). By the 1920's and 1930's, EIS was being employed to gain information about various biological systems, including blood (Fricke, 1925), cell membranes (Cole, 1932), whole sea urchin eggs (Cole, 1928) and contracting muscles (Dubuisson, 1937). The apparatus used at this time was complex and very bulky, with some pieces of equipment such as the Wheatstone bridge, used for measuring resistance and capacitance, taking up an entire room (Fricke, 1925).

In 1947, Randles described a method of fitting impedance data to a model circuit, termed Randles' equivalent circuit (see Section 3.4), which made EIS data obtained from solid interfaces much easier and more robust to interpret (Randles, 1947). Together with other advances in the mathematical processing and modelling of EIS data, the use of EIS became more widespread in the analysis of processes occurring upon surfaces, particularly in the study of corrosion in materials science and bioengineering (Lemaitre et al., 1985).

Whilst an impedimetric biosensor had not yet been developed, other types of electrochemical biosensors were already showing promise. The blood glucose sensor was first marketed commercially in 1975 by the Yellow Springs Instrument Company (Ohio, USA) and there are now over 40 different types of blood glucose sensor on the market (Newman and Turner, 2005). The first capacitance-based affinity sensor emerged in 1986 (Newman et al., 1986) followed by the first impedimetric sensor in 1988 which allowed the detection of small molecules such as acetylcholine (Taylor et al., 1988). In the last two decades, enormous progress has been made in the development of impedimetric biosensors for a diverse range of applications including medicine, pharmaceuticals, environment and anti-terrorism. As discussed in Section 2, great progress has been made in the area of medical biosensing, where impedimetric sensors have been developed to detect a diverse range of analytes within patient samples, including small molecules such as drugs and metabolites, protein biomarkers of disease, nucleic acids, up to entire pathogenic viruses and bacteria. Some of these examples will be presented in further detail in Section 6.

3.2 An overview of impedance

Electrochemical impedance spectroscopy (EIS) is an effective tool for monitoring binding processes occurring at surfaces. In the case of biosensing, EIS is used to detect analyte binding to the surface of bioreceptor-modified electrodes (Katz and Willner, 2003). Due to its high sensitivity, rapid readout and minimal requirement for reagents, impedance spectroscopy is a powerful tool for interrogating biosensors for medical applications (Millner et al., 2012).

As described in Section 2.5.3.3, an impedimetric biosensor converts analyte binding to the sensor surface into an impedance signal, the magnitude of which is proportional to the concentration of analyte. Electrical impedance (Z) is defined as the measure of the opposition that a circuit presents to the passage of a current when a voltage is applied (Chang and Park, 2010). Impedance is a complex term comprising elements of resistance (R; the opposition of the flow of current) and capacitance (C; the storage of charge after the application of a potential).

In order to measure the capacitance and resistance of a biosensor surface, which comprise impedance, the surface is usually incubated in a solution containing electron mediators. Therefore, the model of impedimetric sensing presented herein is based on a system containing solution-based electron mediators. Redox pairs such as $Ru(NH_3)_6^{3+/2+}$ (hexaammineruthenium III/II ions) and the more commonly-used $Fe(CN_6)^{3-/4-}$ (ferricyanide/ferrocyanide) pair allows for the transfer of electrons between these solution-based carriers and the working electrode of the biosensor (Chang and Park, 2010). However, impedance interrogation can also be achieved in the absence of added electron mediators.

Impedimetric biosensors rely upon the binding of analyte to the biorecognition element creating a measurable alteration in conductivity across the biosensor surface. The impedance of any surface, including that of a biosensor, is altered by the deposition of materials upon the sensor surface. Typically, the more bulk material that is present on the surface, the higher the impedance due to (a) an increase in resistance, as the electrons present in solution now have to pass through a thicker layer of material to access the surface and (b) increased capacitance of the bound material, which stores up charge on the sensor surface. However, the relationship is not simple and analyte binding that significantly alters sensor surface nanostructure or chemical nature (e.g. increases hydrophobicity) can lead to a decrease in impedance.

A typical impedimetric biosensor is built upon a solid electrode surface, such as gold, platinum or carbon, as will be discussed in Section 4.1, which conducts charge when connected into a circuit. The transducer surface is usually functionalised, typically by a polymer or self-assembled monolayer (SAM), to allow the attachment of bioreceptors (Section 4.2). This creates an insulating layer on the surface, known as a dielectric, which has both resistive and capacitative components. When the biosensor is incubated in a solution containing electron mediators, the surface presents a certain degree of impedance, or opposition of the electrons from reaching the surface. This can be measured electrically, as described shortly. If analyte is present in the solution, it will bind to the biorecognition element, thereby increasing the amount of material at the sensor surface which alters impedance (Figure 3-1), generally leading to an increase.

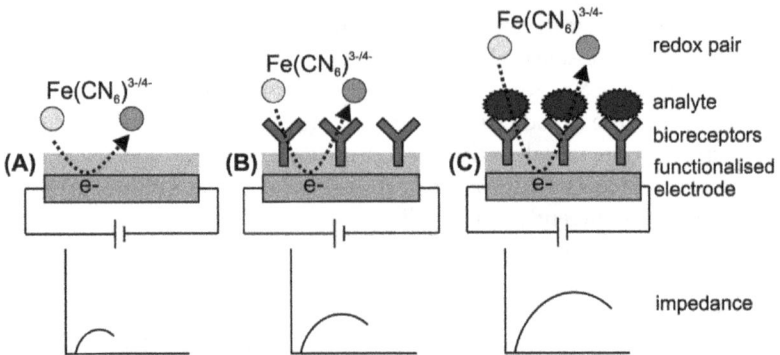

Figure 3-1 Impedance increases with increasing deposition upon a biosensor surface. Increasing deposition of material upon a biosensor surface generally causes an increase in resistance and capacitance which impedes the transfer of electrons between solution-based mediators (such as the redox pair $Fe(CN_6)^{3-/4-}$) and the electrode. Thus, impedance increases from (A) bare transducer surface when (B) bioreceptors are attached to the surface and (C) increases further upon analyte binding to the bioreceptors.

In summary, the electrical impedance of a biosensor surface through which current is flowing typically increases upon analyte binding. The change in impedance is proportional to analyte concentration.

3.3 Principles of impedimetric sensing

In order to obtain data from an impedimetric biosensor system, the biosensor is connected into a circuit to which an alternating potential (V) is applied. This potential is typically a small amplitude sine wave (e.g. ± 10 mV) around a set background voltage. Upon analyte binding, the relationship between the current (I) and voltage alters due to the increased resistance and capacitance of the system.

The impedance of the system is calculated as the ratio between the change in the applied voltage and the change in current, which can be measured when the two are plotted as a sine wave of amplitude vs. time, or phase angle (Figure 3-2). The phase angle or delay arises at each cycle since charging (capacitance) must occur before current can pass.

The impedance of the system is related to both the change in phase (θ) and magnitude ($|Z|$) of the current sine wave relative to the voltage sine wave:

Specifically:
$$|Z| = \frac{V \sin \omega t}{I \sin(\omega\, t\, \theta)}$$

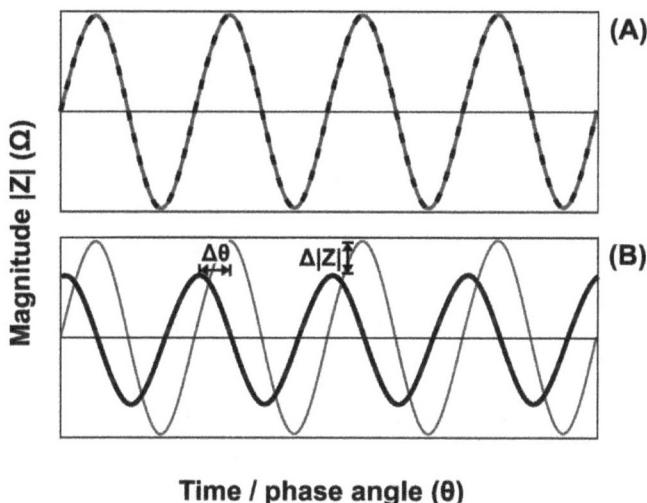

Time / phase angle (θ)

Figure 3-2 Analyte binding to a biosensor surface alters properties of the current with respect to the voltage. Current and voltage can be plotted as sine waves shown as a function of time
or phase angle (θ) and magnitude ($|Z|$). (A) Prior to analyte
binding, the current (black line) and voltage (grey line) are in
phase and of the same magnitude. (B) Analyte binding causes
the current to alter with respect to the voltage. The difference
in phase angle (Δθ) and magnitude ($|Z|$) can be measured and
these parameters both contribute to impedance.

(where $|Z|$ = impedance; V = voltage; I = current; t = time; θ = frequency
of oscillating voltage; θ = phase angle).

Or, more simply: Impedance (Z) = Phase shift (θ) + Magnitude
($|Z|$).

This expression involves the use of a vector term because phase is measured as an angle. Mathematically, these polar coordinates give a value of
impedance. However, these data would be much more useful if presented
as Cartesian coordinates which would allow graphical representation and
analysis. In order to transform the polar data to Cartesian data, the phase
angle is converted into two parameters which comprise the total impedance
in an electrical circuit; Z' (the "real" impedance component, similar to resistance) and Z" (the "imaginary" impedance component, similar to capacitance) (Figure 3-3).

The resulting real and imaginary components of impedance may now
be represented as a Nyquist plot, from which useful impedance data can
be extracted. The take-home message from this section is that impedance

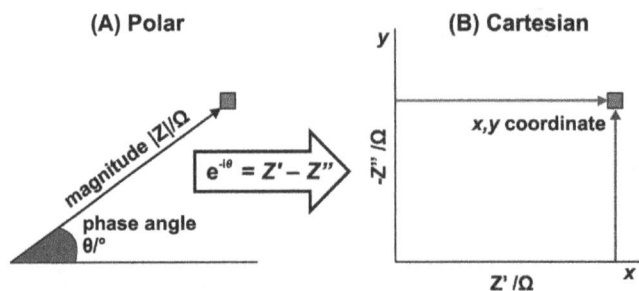

Figure 3-3 Conversion of impedance data from polar to Cartesian coordinates. The phase angle is related to the real (Z') and imaginary (Z") components of impedance by the equation: $e^{-i\theta} = Z' - Z''$, where e = natural logarithm, i = square root of -1 (imaginary number), θ = the phase angle in degrees, Z' = real impedance in ohms and Z" = imaginary impedance in Ohms.

comprises of two parameters; resistance and capacitance, which change upon analyte binding to a biosensor surface, and which can be measured and depicted graphically.

3.4 Presenting and analysing impedance data

A Nyquist plot (Figure 3-4a) is the usual way of representing impedance data. Electrochemical impedance spectroscopy (EIS) measurements are obtained by applying an alternating voltage at a range of frequencies; a typical range for biosensor studies being 25 kHz down to 0.25 Hz (Billah et al., 2008; Caygill et al., 2012). The real (Z') and imaginary (Z") components of impedance are plotted on a Nyquist plot.

The features of the Nyquist plot can be related to components of the common model for fitting impedance data, the Randles' equivalent circuit (Figure 3-4b). The model arises from so-called Faradaic or non-Faradaic measurements. In a Faradaic process, charge is transferred from the solution across the biosensor surface to the electrode. In a non-Faradaic process, current flows without any charge transfer; instead, the charge accumulates on the capacitative components of the surface. Thus, non-Faradaic sensors are considered to be capacitive, whereas the model of impedimetric sensors described herein refers to Faradaic sensors which operate in the presence of redox probes to facilitate charge transfer.

The model of an impedimetric biosensor as part of a circuit includes solution resistance (R_s), double-layer capacitance (C_{dl}), charge transfer resistance (R_{ct}) and, in the case of Faradaic sensors, Warburg impedance (W). The solution resistance arises from the conductance of ions in the bulk solution and therefore remains the same pre- and post-analyte binding. The double-layer capacitance arises from the local build-up of charge upon the

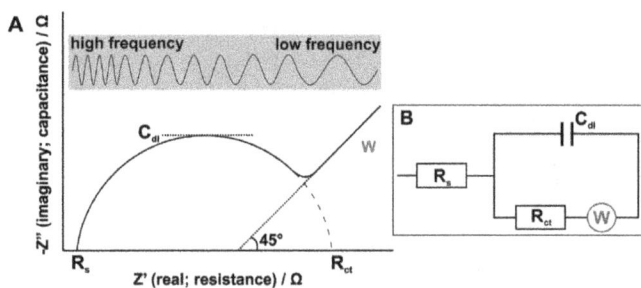

Figure 3-4 Schematic of a Nyquist plot alongside a Randles' equivalent circuit. (A) A Nyquist plot shows the real (or resistive) and imaginary (or capacitive) elements of impedance. The resistance of the solution (R_s) and the charge transfer resistance (R_{ct}) can be calculated from the x-axis intercepts. The maximum double-layer capacitance (C_{dl}) is represented by the height of the semi-circle. Double layer capacitance refers to the dielectric boundary which is formed by the insulating layers on the biosensor surface, which separate the electrode from the solution. (B) Randles' equivalent circuit is a way of modelling the system. The Warburg impedance (W), only observed in the case of Faradaic sensors, represents mass transfer diffusion effects which come into effect at low frequencies.

sensor surface, which can be altered by analyte binding. The charge transfer resistance is due to the insulating layer on the sensor surface which hinders the transfer of electrons from the solution to the electrode; this can also be affected by analyte binding. The Warburg impedance arises because the electron mediators diffuse to the sensor surface at a finite rate; this is not usually affected by analyte binding. Thus, the parameters that change upon analyte binding are the double-layer capacitance (C_{dl}) and the charge-transfer resistance (R_{ct}), and these are often plotted against analyte concentration for biosensor calibration plots.

The semi-circular trace of the Nyquist plot reflects the behaviour of the sensor surface at different frequencies of alternating voltage. At high frequency, the oscillation in current is too fast for electron transfer to take place between the solution-based electron mediators and the sensor surface. Therefore, at high frequency, any electron transfer is occurring on top of, but not through, the sensor surface. The x-axis intercept represents this parameter, known as the solution resistance (R_s) which is not dependent upon sensor surface parameters and therefore remains constant regardless of analyte binding.

In contrast, at low frequencies, there is time for electrons to be transferred from the solution-based mediators through the insulating layers on

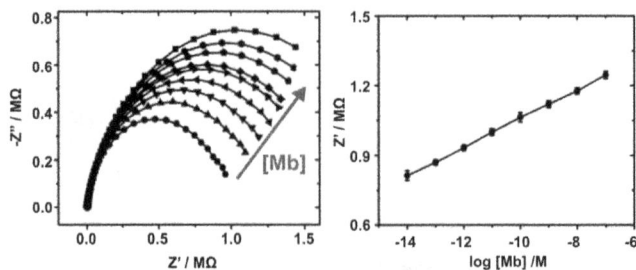

Figure 3-5 Real data obtained from myoglobin impedimetric affinity sensors. (A) Nyquist plots of impedance spectra recorded with myoglobin (Mb) immunosensors in the presence of increasing concentration of myoglobin (10^{-14} to 10^{-7} M, increasing in the direction of the arrow). (B) Calibration curve for myoglobin immunosensor. The real component of impedance (Z') is plotted against concentration of myoglobin at a frequency of 0.25 Hz. Data adapted from (Billah et al., 2010).

the biosensor surface through to the electrode beneath. Here, the electrons are subjected to capacitance as well as resistance. The semi-circle can be extrapolated back down to the x-axis to obtain a value which consists of solution resistance (R_s) and charge-transfer resistance (R_{ct}).

Increasing deposition of material, which occurs when analyte binds to the sensor surface, typically retards electron transfer across the surface (although, as discussed earlier, this is not always the case). This generates an increase in both capacitance and resistance, which is reflected in larger semicircular traces on a Nyquist plot (Figure 3-5a). The impedance (typically Z' or R_{ct}) can be plotted against analyte concentration (Figure 3-5b). In an ideal system, there would be a linear relationship between impedance and analyte concentration.

4. Fabrication of impedimetric biosensors

The fundamental principle behind biosensing is the conversion of a specific biorecognition event into a measurable signal. As described in Section 2.4, a wide variety of bioreceptors can be employed to detect a diverse range of analytes. The signal may be transduced in a variety of ways; electrochemically, optically or mechanically. In the case of impedimetric sensors, a high-affinity interaction between the analyte and bioreceptors, which are often antibodies, gives rise to alterations in the electrical properties of the sensor surface. Therefore, the layer-by-layer construction and the nanostructure of the biosensor surface must be finely-tuned in order to achieve specific, measurable and robust analyte detection within complex patient samples.

The layer-by-layer construction of impedimetric biosensors begins with a conducting surface onto which an alternating voltage is imposed and through which an alternating current is induced. This is typically a noble metal or carbon electrode that forms part of a small, disposable sensor chip. The transducer surface is functionalised and bioreceptors are attached in order to produce an affinity biosensing platform. Ultimately, the system should lend itself to manufacture on a commercial scale and the sensor surface chemistry should be optimised to prevent the non-specific attachment of other substances in the test samples.

Furthermore, in the case of implantable biosensors for continuous patient monitoring, additional consideration must be given to the materials used for the entire sensor device. The materials need to be non-toxic and should not provoke any immune response; bio-fouling should be minimised by appropriate insulation and the electrodes should be corrosion-resistant under *in vivo* conditions (Geddes and Roeder, 2003). For invasive or implantable monitoring in clinical situations, as well as considering the immunogenicity and toxicity of the materials, the device must be very small and any readout device or power supply must also be miniaturised appropriately (Grieshaber et al., 2008). For all commercially-available biosensors, particularly those intended for use in warm climates, the stability of the materials is extremely important in order to extend the shelf-life and to ensure a robust readout.

The design and optimisation of the electrodes, functionalisation of the transducer surface and bioreceptor selection and attachment are, therefore, all critical parameters that are essential to the success of the biosensor system. In this section, we will consider these practical aspects of building impedimetric biosensors for medical applications.

4.1 Electrode design and materials

Three different electrodes are typically required to measure impedance (Daniels and Pourmand, 2007) (Figure 4-1). The working electrode is functionalised with bioreceptors to form the biosensing element. The current at the working electrode is then measured with respect to a reference electrode,

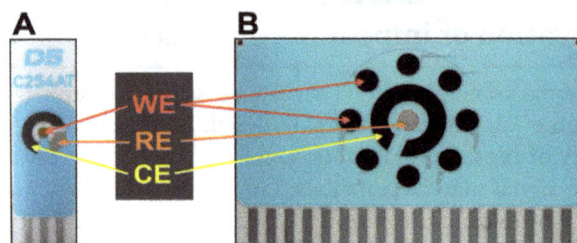

Figure 4-1 Commercially available electrode chips with alternative layouts for impedimetric biosensors. Screen-printed electrode chips from DropSens S.L. (Oviedo, Spain). Electrodes may have (A) one or (B) multiple working electrodes (WE), along with a reference (RE) and counter (CE). Images provided by DropSens S.L. (Oviedo, Spain).

which maintains a fixed potential between the metal contact and the solution. Finally, a counter electrode supplies current to the solution in order to maintain the desired potential between the electrodes and the solution.

The design of the electrodes upon the sensor chip can vary according to the materials used, the nature of the analyte and test samples and according to the way in which the chip will plug in to the integrated platform. Sensor chips may also be designed with array of multiple working electrodes for multi-analyte detection (Komarova et al., 2010; Yu et al., 2006) (Figure 4-1).

4.1.1 Base electrode materials

The working electrodes must be highly conductive with low bioactivity and low binding of biomolecules. Furthermore, the ability to manipulate the materials to form geometric shapes on an industrial scale is important. This has generally limited the choice of electrode materials to noble metals such as gold (Au) and platinum (Pt) as well as carbon (Huang and Suni, 2008).

Gold has proven to be a popular choice of electrode material for EIS since the first reported impedimetric biosensor in 1988 utilised gold electrodes that had been deposited onto quartz chips (Taylor et al., 1988). In spite of its relatively high cost, gold offers several advantages including high conductivity and the ability to achieve almost atomically-flat surfaces depending upon the deposition method. As will be discussed in Section 4.2, surface flatness can be important in transducer surface functionalisation. Gold also permits the direct attachment of molecules through the chemi-adsorption of thiol groups (Millner et al., 2009) (Section 4.2). Gold electrodes have been employed to detect a wide and diverse range of clinically-relevant analytes, including pathogens (Caygill et al., 2012; Lu et al., 2013), disease biomarkers (Billah et al., 2008; Hu et al., 2013), indicators of traumatic injury (Arya et al., 2013) and small molecules (Chauhan et al., 2013; Chullasat et al., 2011).

Platinum possesses similar properties to gold, although it is generally more expensive, and platinum electrodes have also been employed in the fabrication of impedimetric biosensors for medical applications (Liu et al., 2009; Peh and Li, 2013). Gold and platinum, as heavier metals, are preferable for implantable biosensor devices as their increased absorption of x-rays facilitates their radiographic visualisation and monitoring, as well as their resistance to corrosion (Geddes and Roeder, 2003). As gold and platinum are corrosion-resistant, this also facilitates the re-use of a biosensor chip by acidic regeneration buffers (Radi et al., 2005), which are described in Section 5.4.

Various forms of carbon have been used as working electrodes in impedimetric biosensors for medical applications (Alizadeh and Akbari, 2013; Barton et al., 2009; Lee et al., 2008). Carbon is much cheaper than gold or platinum and, therefore, makes the end product much more attractive commercially. Carbon paste, comprising of powdered carbon in combination with a polymer, has been one of the most popular electrode materials since its invention in 1958 (Svancara et al., 2009). The powdered carbon may be in a variety of forms, including soot, charcoal, glassy carbon and, more recently, carbon nanofibres and nanotubes. Glassy carbon electrodes (GCEs) are often used in electrochemical biosensors due to their low electrical resistance and high stability (Kashefi-Kheyrabadi and Mehrgardi, 2012).

Pyrolytic carbon is an alternative to carbon paste and is formed when carbon is heated to a high temperature and then left to crystallise. Pyrolytic carbon is already used to coat artificial joint replacements so would be particularly suitable for implantable biosensors due to its biocompatibility (Tägil et al., 2013). Pyrolytic graphite electrodes (PGEs) have been used in several impedimetric biosensors and offer lower capacitance and lower background current than glassy carbon electrodes (Lee et al., 2007).

Graphene, discovered in 2004, is an atomically thick layer of carbon in which the atoms are arranged in a honeycomb-like lattice (Novoselov et al., 2004). Graphene oxide, a precursor of graphene, is of particular interest in medical biosensing and *in vivo* applications due to its high electrical conductivity, good biocompatibility, large surface area, low cost, facile synthesis and ease with which it can be functionalised and deposited upon solid surfaces (Bonanni et al., 2012; Kim et al., 2011). Graphene oxide offers the advantage that nucleic acids can be adsorbed directly onto its surface. This has allowed for use of single-stranded DNA as bioreceptors to detect complementary DNA sequences, such as fragments of the human immunodeficiency virus (HIV) genome (Hu et al., 2012) and immobilised DNA aptamers to detect proteins such as thrombin (Loo et al., 2012).

Various other materials have also been utilised in the fabrication of electrodes for impedimetric biosensors, either as the base layer material or as a surface modification. A wide range of metals, including titanium, chromium, rhodium, palladium, nickel, copper, aluminium, iron and silver have either been used alone or alloyed with gold or platinum (Geddes and

Roeder, 2003; Ohno et al., 2013). However, not all of these are suitable for clinical biosensors, in particular implantable devices, as some metals such as nickel are highly immunogenic.

4.1.2 Electrode modifications at the nano-scale

Impedimetric biosensors are often required to detect very low concentrations of analyte within complex matrices, such as blood, which contain many other interferents. In order to increase the signal generated upon analyte binding to the biorecognition element, the electrode surface can be modified using nanomaterials with the effect of increasing the surface area for receptor-analyte interactions.

Nanomaterials are generally defined as each unit having a length of between 1–100 nm, with the material in this form having different chemical, electronic and mechanical properties from the bulk substance. A range of nanomaterials have been employed in the design of electrodes for impedimetric biosensors (Suni, 2008); these include nanoparticles (spheres) and nanotubes (rods) of gold, silver, copper, carbon and other materials (Figure 4-2). Gold nanoparticles (AuNPs) and carbon nanotubes (CNTs) are the most widely used due to their high conductivity, biocompatibility and commercial availability.

Impedance sensor electrodes modified by AuNPs generally take one of two approaches; either the AuNPs are deposited as one flat layer on the top of a conducting electrode, or they are used in conjunction with a cross-linking reagent to form a three-dimensional colloidal network. Gold nanoparticles are commonly used in impedimetric biosensing. For example, using colloidal AuNPs in the construction of a Hepatitis B biosensor was shown to increase the attachment of bioreceptors and gave a detection limit of 50 ng/l (Wang et al., 2004). Gold nanoparticles have also been used in impedimetric DNA sensors to increase attachment of probe DNA and to boost sensitivity as compared to bare gold electrodes (Zhang et al., 2008).

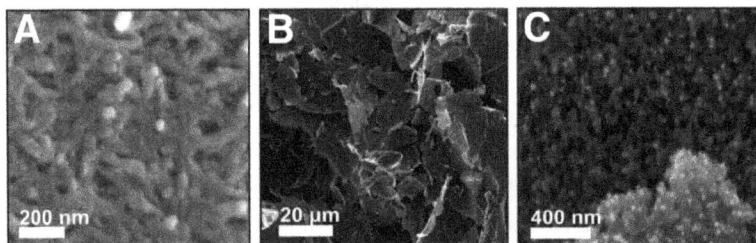

Figure 4-2 Nano-modifications to electrode surfaces. Scanning electron microscopy images of (A) carbon nanotubes, (B) graphene platelets and (C) gold nanoparticles, all deposited upon screen-printed electrodes. Images provided by DropSens S.L. (Oviedo, Spain).

CNTs, hollow tubes formed from rolled-up sheets of graphene with either single walls (SWCNT) or multiple walls (MWCNT), have been used in various impedimetric sensor platforms to improve sensitivity and signal. They exhibit high tensile strength, conductivity and chemical stability and are easily functionalised with biomolecules such as proteins and nucleic acids (De Volder et al., 2013). Recently, an impedimetric biosensor for the detection of a biomarker of liver cancer, alpha-foetoprotein (AFP), was developed by layering bioreceptor-functionalised SWCNTs upon carbon electrodes (Yang et al., 2013a). This approach has brought the limit of detection down to 0.1 ng/l of protein in serum-containing patient samples. The attomolar detection of target DNA was achieved by the attachment of probe DNA onto polymer-functionalised SWCNTs that had been deposited onto platinum electrodes (Kurkina et al., 2011). Similarly, MWCNT-modified gold electrodes have been used for impedimetric DNA sensing at picomolar concentrations (Yang et al., 2013b). Thus, electrode surface nano-modification with CNTs can facilitate high changes in impedance in response to a very low degree of analyte binding.

Nanowires (NW) are an emerging class of functional materials that enhance the surface properties of electrodes for medical impedimetric biosensing (Patolsky et al., 2006). These nanometre-thick wires may be synthesised from the bottom-up (grown from smaller particles) or from the top-down (by fragmenting a larger piece of material). Both conducting and semi-conducting nanowires have been utilised in impedimetric biosensing; examples include gold nanowires (Ramulu et al., 2013) and gallium nitride nanowires for DNA sensing (Sahoo et al., 2013), titanium oxide nanowires for bacterial sensing (Wang et al., 2008) and silicon nanowires for the detection of hepatitis B and liver cancer biomarkers (Ivanov et al., 2012). It should be noted that the type of nano-modification must be optimised for the particular type of electrode, as this is not a one-size-fits-all principle (Arya et al., 2013).

4.1.3 Electrode manufacture

A wide range of materials can be utilised successfully in the design of working electrodes for laboratory-based impedimetric biosensing. In order to translate these research-stage biosensors into a commercially-available product, the electrodes must be amenable to miniaturisation and mass manufacture, and the process must be cost-effective.

In the 1960s and 1970s, when biosensing was an emerging field following the development of the glucose sensor in 1962, a biosensor was a bulky device, similar to a pH electrode (Luong et al., 2008). The first commercially-available glucose meter, the YSI Blood Glucose Analyzer Model 23A (Yellowsprings Instruments, 1975), was a large piece of equipment measuring 40 cm in width. The electrode consisted of a buffer-filled probe with a platinum working electrode at the "test" end. The probe contained a membrane with immobilised enzyme as bioreceptor, which was re-used up until a non-linear calibration was observed, at which point it was replaced. The membrane had to be checked every day and sensor maintenance and cleaning was vital to keep the system functioning.

Much progress has been made since the launch of this early system in order to develop disposable electrodes that operate within small, portable systems without the need for sensor cleaning and calibration (Wang, 2001). Screen-printing, otherwise known as thick-film technology, was adapted from the microelectronic industry in the 1990s and offers high volume production of disposable, low-cost electrodes for electrochemical biosensing. The application of screen-printing to the fabrication of impedimetric biosensors in recent years has enabled the mass production of screen-printed electrodes (SPEs) which are inexpensive, reproducible, sensitive, easily modified and require little or no pre-treatment before use (Metters et al., 2011). SPEs represent the current state-of-the-art for electrochemical sensors used *in situ* because of their quick and sensitive response, ability to operate at ambient temperatures, linear output and low power requirements.

In screen printing technology, a conductive ink paste is rolled onto a solid substrate, usually plastic or ceramic, through a mesh stencil and then hardened, with an insulating ink applied to separate the conductive track from the electrodes (Zhang, 2000). Silver ink is often employed for the conducting track, whereas carbon or gold is most common for the working electrodes. Screen-printing offers enormous versatility in terms of the materials and modifications that can be employed in electrode deposition. The composition of the printing ink can be altered easily by the addition of metals, polymers, complexing agents etc. and it is also possible to further modify the electrodes by the deposition of metal films, polymers etc. upon the surface (Renedo et al., 2007). In this automated process, many thousands of electrodes can be produced robotically in the same batch, thereby lowering costs and decreasing electrode-to-electrode variability, which is currently an important issue for impedimetric sensors. The resulting sensor chips are disposable in nature and, due to their small size, reduce the sample volume to microlitres. This is extremely useful in biomedical sensing whereby a finger prick can be utilised instead of a blood test, reducing time, discomfort to the patient and negating the need for trained medical personnel to obtain the sample.

These "lab-on-a-chip" devices are much more amenable to clinical applications and nowadays the majority of commercially available biosensors utilise disposable, screen-printed electrodes (Wang, 2001). Various companies around the world now offer both standard and custom-designed screen-printed electrodes, including Metrohm USA Inc. (USA), DropSens S.L. (Spain), Gwent Sensors Ltd. (UK), Bio-Logic SAS (France), Kanichi Research Ltd. (UK), BVT Technologies Ltd. (Czech Republic) and Quasense Company Ltd. (Thailand).

4.2 Bioreceptor tethering to transducer surfaces

4.2.1 Transducer surface functionalisation

The transducer surface of an impedimetric biosensor, as discussed in Section 2.5, forms the foundation upon which the sensing platform is constructed by the attachment of bioreceptors. Although in certain cases the bioreceptors can be deposited directly onto the electrode by chemisorption (van Noort and Mandenius, 2000), the metal surface has a high surface energy which may impair the function of the biorecognition component (Lu et al., 1996). Functionalising the transducer surface creates a spatial barrier preventing direct contact between the metal surface and the bioreceptors and also provides chemical moieties which enable the stable attachment of the bioreceptors to the electrode surface (Hermanson, 2008; Millner et al., 2009).

There are several common methods of transducer surface functionalisation and bioreceptor tethering; these are matrix entrapment and self-assembled monolayer (SAM)- or polymer- based attachment (Figure 4-3). The tethering strategy is largely determined by the choice of electrode material, the chemical nature and surface topology of the electrode and the size and type of analyte.

4.2.1.1 Matrix entrapment.

One of the first methods employed to attach bioreceptors to a transducer surface was matrix entrapment (Figure 4-3a). The bioreceptors are combined with a mixture of monomers or oligomers of a long-chain molecule such as Prussian Blue (Pchelintsev et al., 2009) or an electrostatic polystyrene sulfonate/poly(diallyldimethylammonium chloride) sandwich (Millner et al., 2009) which can be either electrostatically (Pchelintsev et al., 2009) or electrochemically (Gerard et al., 2002) deposited onto the surface of the electrode. This results in a surface matrix with the bioreceptors distributed throughout in a non-oriented fashion (Ramanavicius et al., 1999; Trojanowicz and Krawczyński vel Krawczyk, 1995). Although this means that the signal may be quickly relayed to the transducer surface, the matrix can hinder analyte reaching and binding to the bioreceptors, which hampers signal generation (Lu et al., 1996). Maximum signal is crucial for sensing very small quantities of analyte within patient samples, so this may represent a significant drawback in using matrix entrapment methods for the construction of medical biosensors, except for metabolites of which the concentration is typically fairly high e.g. urea, lactate, glucose etc.

4.2.1.2 Self-assembled monolayer (SAM)-based attachment.

If the electrode surface is particularly smooth, it may allow the construction of self-assembled monolayers (SAMs) (Figure 4-3b). SAMs are an organised array of closely-packed molecules which spontaneously bind to a surface in a predictable and ordered fashion. As the properties of the SAM

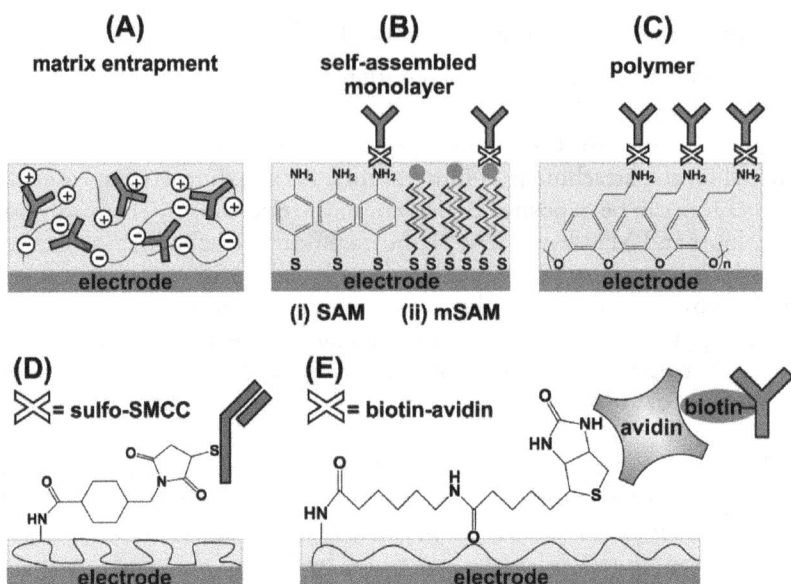

Figure 4-3 Schematic of common transducer surface function-alisation and bioreceptor tethering strategies. The three common methods of transducer surface functionalisation are (A) matrix entrapment; (B) self-assembled monolayers (SAMs) such as (i) 4-aminothiophenol, 4-ATP, or (ii) mixed-SAMs (mSAMs) such as mercaptohexadecanoic acid(MHDA)/biotin-caproyl-DPPE, and (C) polymers such as polytyramine. Common methods of tethering bioreceptors to SAMs and polymers include (D) co-valent cross-linking of pendant surface amines *via* sulfo-SMCC to sulfhydryl groups, such as are present on reductively-cleaved half-antibodies, and (E) the non-covalent biotin-avidin inter-action, whereby the transducer surface and bioreceptors are biotinylated (e.g. using NHS-biotin) and linked together by a tetrameric avidin derivative which has multiple biotin binding sites. The linker (X) shown in schematics B and C can include the structures indicated in schematics D and E.

can be tuned by changing their molecular composition, they are a good can-didate for biosensor construction and indeed many impedimetric biosensor systems have been developed based on SAMs (Wink et al., 1997).

The SAM molecules typically have a different functional group at either end, such that one end of the molecule attaches to the electrode surface and the other end enables tethering of the bioreceptors. SAMs generally comprise of either long chain molecules such as mercaptohexadecanoic acid (MHDA) (Billah et al., 2010; Rodgers et al., 2010) and other fatty acids (Xu et al., 2013), such that the resulting SAMs resemble one leaflet of a cell membrane, or alternatively are small, aromatic molecules such as 4-aminothiophenol (4-ATP) (Conroy et al., 2010) (Figure 4-3bi). Mixed SAMs (mSAMs) result from a mixture of two or more molecular components that are combined to create the surface layer. The different components are held together by non-covalent forces, as in the example of MHDA:biotin-caproyl-1,2-dipalmitoyl-*sn*-glycero-3-phosphoethanolamine (biotin-caproyl-DPPE) based mSAMs (Figure 4-3bii) (Billah et al., 2010) and 11-aminoundecane-thiol in array with 6-mercaptohexadecanol (Mantzila et al., 2008).

In SAM-based transducer surface functionalisation, the first consideration is the electrode surface-tethering end, known commonly as the head group. This interaction between the SAM molecule and the electrode surface must be strong as it forms the basis for biosensor assembly. One interaction which is commonly employed is the gold-thiol (Au-SH) interaction (Bain et al., 1989; Fenter et al., 1994). This spontaneously forms a dative bond between the electrode surface and the SAM which is particularly strong (Hakkinen, 2012). Although other metals can be used for the electrode surface, gold is often preferred due to the diverse array of functionalisation routes. Platinum has been utilised for certain systems (Prodromidis, 2010) whereas silver is often a poor substrate for SAM adhesion and so a binding mediator is required (Zhang et al., 1998).

As well as thiol-mediated SAM tethering, another reaction which is commonly exploited is silanisation (Hermanson, 2008). These reactions involve the displacement of surface hydroxyl groups by reactive alkoxyl groups. As many metals have a surface hydroxide layer this allows for non-gold electrodes to be functionalised. Some popular molecules used are aminopropyl-triethoxy silane (APTES), and glycidoxypropyl dimethylethoxysilane (GPMES) (Prodromidis, 2010). Silanisation is beneficial for the development of impedimetric sensors as it enables other cheaper metals to be used in electrode manufacture, lowering the cost of the final device.

Although the head groups are often tethered covalently, as in the examples of thiolated SAMs on a gold surface, SAMs and particularly mSAMs benefit from intermolecular stabilisation by hydrophobic interactions and other non-covalent forces (Flynn et al., 2003). This means that SAMs are relatively weak as a transducer surface platform and so may only be suitable for the detection of smaller analytes such as metabolites and proteins (Xu et al., 2013) rather than larger analytes such as bacteria and viruses, which can damage the integrity of the SAM.

Once the SAM has been assembled upon the electrode surface, bioreceptors may be tethered to the SAM either directly or *via* a bifunctional chemical cross-linker (Figure 4-3d&e) (Hermanson, 2008). Although SAMs are highly ordered molecular assemblies, generally the bioreceptors attach in a random fashion, as the tethering group may be located anywhere on the surface of the bioreceptor.

Although SAMs provide a substrate which is spatially organised for bioreceptor tethering, they may still encounter problems with steric hindrance (Zheng and Du, 2013). This can reduce greatly the signal generated by analyte binding (Lu et al., 1996). In the context of biosensors, steric hindrance is defined as the spatial restrictions which impede the analyte from reaching the sensor surface or the bioreceptor itself (Piro et al., 2007). This can be due to active site occlusion where the bioreceptor's ligand binding domain is bound to a substrate facing the surface.

The electrode material is a key consideration when selecting the functionalisation and bioreceptor tethering strategy. Whilst SAMs generate very clean data, they can only be constructed upon very flat surfaces where the roughness is less than the depth of the SAM. This is typically a few nanometres; any greater roughness tends to bring about defects in the SAM and compromise the electrochemical signal (Luppa et al., 2001). This is particularly a problem for medical immunosensors where the analyte may be relatively large, such as a bacterium or virus. Also the highly specialised surfaces required for SAM construction tend to be prohibitively expensive for general use in point-of-care diagnostics.

The nature of the analyte itself is highly influential in defining the sensor construction route as larger analytes such as bacteria and whole cells tend to disrupt SAMs (Hong et al., 2009). To counteract this, more robust tethering substrates must be used. This is due to the relative sizes of the forces holding the transducer together and those required to bind the analyte. If the analyte is large with multiple binding epitopes, a SAM is unsuitable as binding may induce fractures in the SAM, thus interrupting the signal. To circumvent this problem, systems have been demonstrated in which the transducer surface is functionalised with a covalently bound matrix which cannot be deformed easily. Polymers are a particularly useful route of achieving this and many conducting polymers have been employed in a range of electrochemical biosensor systems (Gerard et al., 2002). They have been demonstrated to generate good signal when bioreceptors are bound creating a very robust sensor construct (Caygill et al., 2012).

4.2.1.3 Polymer-based attachment.

Conducting polymers (Table 4-1 and Figure 4-3c) are a widely used substrate for the construction of impedimetric biosensors, as they are relatively inexpensive and provide a robust surface for the attachment of bioreceptors. Polymers are typically attached to the electrode surface by the process of electropolymerisation, which is carried out by cyclic voltammetry with the electrode immersed in monomer solution (Caygill et al., 2012; Raffa et al., 2006). Electropolymerisation is a selective process which can be precisely controlled to generate a well-defined polymer substrate. One drawback is the time taken and instrumentation necessary for deposition; for this reason, contemporary research is exploring the possibility of printing pre-made polymers onto electrodes (Reddy et al., 2011) for the creation of screen-printed biosensors. This has the benefit of scalability for mass production and, therefore, low cost.

Table 4-1 Examples of commonly used polymers used in transducer surface functionalisation.

Monomer	Polymer
aniline	polyaniline
pyrrole	polypyrrole
tyramine	polytyramine
thiophene	polythiophene

The polymers used in transducer surface functionalisation are conducting in order to permit charge transfer to allow for the transduction of electrochemical signals. Many conducting polymers are based on benzene ring-containing or phenolic compounds due to the large field of delocalised electrons permitting conduction. One polymer which has been widely reported on the construction of impedimetric biosensors is polyaniline. Successful examples include sensors for whole viruses (Caygill et al., 2012) as well as sensors for cardiac drugs (Barton et al., 2009). Polyaniline is a good candidate as it can be easily polymerised and incorporates many pendant amine groups to facilitate the attachment of bioreceptors. Other polymers have also been used such as polytyramine (Pournaras et al., 2008) (Table 4-1); polytyramine is useful as it has fewer available oxidation states so may create a more electronically stable substrate than polyaniline (Scouten et al., 1995). This chemical diversity is a key advantage of polymers, as the properties of the transducer surface can be tuned for optimum signal generation and minimum background binding of any particular analyte by varying the polymer composition.

4.2.2 Directed *vs* undirected orientation of bioreceptors

Where the bioreceptors are attached to the transducer surface in a non-directed way, an associated reduction in biosensor signal has been observed (Lu et al., 1996). By orienting the bioreceptors on the sensor surface to optimise analyte binding, we can improve the signal from the biosensor (Bonroy et al., 2006). These sensors give a greater analyte-specific signal, underlining the importance of the nano-architecture of the biosensor interface.

To achieve oriented attachment, bioreceptors can be manipulated on a molecular level in various ways (Scognamiglio, 2013). One recently used method is the generation of half-antibodies, shown tethered to the transducer surface in Figure 4-3d (Billah et al., 2010; Caygill et al., 2012). Half-antibodies can produced by exposing antibodies to a mild reducing agent, such as 2-mercaptoethylamine or tris (2-carboxyethyl)phosphine (TCEP), which is strong enough to break only the disulfide bond between the two Fc regions, leaving the other structural disulfide bonds (with marginally higher bond energies) intact. The resulting half-antibodies now present unique cysteine residues, containing pendant –SH groups, which permit their attachment to the transducer surface such that the ligand-binding region is facing away from the sensor surface and towards the sampling region. Oriented half-antibodies have been shown to increase the sensitivity of analyte detection compared with non-oriented full antibodies (Billah et al., 2008; Billah et al., 2010). In the case of non-antibody bioreceptors, such as nucleic acids or non-antibody binding proteins, unique functional groups can be engineered biochemically into a specifically chosen site that is distal from the ligand-binding region (Gilbreth and Koide, 2012).

4.2.3 Bioreceptor tethering to the transducer surface

Once the electrode surface has been suitably functionalised, the bioreceptors must be attached in a permanent and robust manner. There are a number of strategies for bioreceptor immobilisation that are routinely exploited, all of which are procedurally simple and which occur under mild conditions. Generally, bioreceptors are tethered to the transducer surface either covalently or using the high-affinity biotin-avidin interaction (Figure 4-3d&e).

4.2.3.1 Covalent cross-linking.

There are a number of different strategies for the chemical cross-linking of bioreceptors onto the transducer surface, the choice of which depends upon the available functional groups presented by both.

The formation of disulphide bonds (-S-S-) is particularly useful as it is an entropically-driven reaction which forms covalent bonds. The reaction requires no extremes in pH temperature or time. Many proteins contain cysteine residues which present thiol (-SH) functional groups, offering an easy pathway to tether them to the biosensor surface assembly. Common cross-linkers such as NHS, EDC and succinimidyl-(N-maleimidomethyl) cyclohexane-1-carboxylate (SMCC) also provide methods to covalently cross-link different components (Hermanson, 2008), as they react with -NH$_2$, -COOH and -SH groups, respectively. Figure 4-3d shows the covalent attachment of thiol-containing half-antibodies onto transducer surface amine groups using the hetero-bifunctional cross-linker, sulfo-SMCC.

4.2.3.2 Biotin-avidin interaction.

One of the most widely exploited interactions is the biotin:avidin interaction. This is a protein:ligand interaction between biotin, alternatively known as Vitamin H, and a complementary binding protein; avidin (Hermanson, 2008). This very high affinity interaction is non-covalent but has a dissociation constant (K_d) of ~10^{-15}M, making it a very stable reaction to exploit in the layer-by-layer construction of biosensors (Figure 4-3b).

The biotin-avidin interaction is widely utilised and a broad range of reactive biotin derivatives are commercially available which react with specific functional groups. These include N-hydroxysuccinimide (NHS)-biotin (reacts with –NH$_2$), maleimide-biotin (reacts with –SH) and the ethyl-dimethylaminopropylcarbodiimide (EDC)-mediated reaction with biotin hydrazide (reacts with –COOH). These reagents allow biotinylation to occur spontaneously under mild conditions in standard buffers, which helps to retain the structure, and hence the bioactivity, of the bioreceptor and maintains the properties of the SAM or polymer matrix. Once the transducer surface and the bioreceptors have been biotinylated they can then be biologically cross-linked by incubation in the presence of an avidin derivative.

4.2.4 Biosensor surface analysis and optimisation techniques

The steps involved in the layer-by-layer construction of a biosensor, which entail the functionalisation of the electrode surface prior to bioreceptor tethering, are absolutely critical to the construction of impedimetric biosensors. These biosensor construction stages can be monitored and optimised using various surface analysis methods (Guan et al., 2004). Not only do these provide confirmation of the various surface modifications but they assist in the optimisation of biosensor fabrication (Tiefenauer and Ros, 2002).

As well as using electrochemical techniques such as cyclic voltammetry and EIS for biosensor interrogation, they are also useful tools for the monitoring and validation of each step of sensor construction, from transducer functionalisation to bioreceptor attachment. Cyclic voltammetry is a useful technique for probing the transducer surface chemistry as the shape of the voltammogram, as well as the position of any redox peaks, provides information about the conducting or insulating nature of the surface (Pchelintsev and Millner, 2008; Rodgers et al., 2010). EIS measurements taken during biosensor construction also serve to verify the deposition of material upon the sensor surface, although they do not provide any information about the chemical properties or the activity of the surface (Corry et al., 2003).

Atomic force microscopy (AFM) is employed widely to confirm and to analyse the successful assembly of biosensor surfaces (Corry et al., 2003; Huang et al., 2010). This is a mechanical method in which a cantilever with an atomic-size tip probes the surface of the sample. The movement of the cantilever mirrors the surface topology and this movement is transformed into a 3D picture of the sensor surface. AFM is a useful scanning method which can provide surface detail at the nanometre scale and which can be used to analyse both wet and dry samples. However, it is associated with high costs and is a labour intensive method to acquire high resolution images. It can also give information about the material properties of the surface such as the elastic properties at the surface, again helping to determine the difference between "hard" transducer surfaces and any softer deposited biosensor layers. This provides further confirmation of biosensor assembly (Caygill et al., 2012). AFM has also been used to characterise the proteins on the transducer surface, assessing distribution and any alterations that may have occurred upon protein binding (Billah et al., 2010).

Scanning electron microscopy (SEM) is a more direct imaging technique in which a beam of electrons is used in place of light to generate high resolution images of a surface. One drawback of this is that the electron beam requires a vacuum so only dry samples can be analysed. Therefore, when considering the role of water in the hydration of biomolecules, any images acquired may not be representative of the biomolecules in their natural, hydrated state. Although SEM provides high resolution images, it cannot give any information on structural or mechanical properties. For comparing surfaces, however, it is a highly effective tool for the comparison of different transducers. For instance, SEM can be used to compare different electrode materials, preparation and cleaning techniques and to explore the nano-architecture of the sensor surface (Kadara et al., 2009), all of which assists in the construction and optimisation of impedimetric biosensors (Yun et al., 2007).

Quartz crystal microbalance (QCM) analysis is also used to monitor biosensor construction. Here, a piezoelectric crystal is subjected to an oscillating voltage which establishes a harmonic frequency. As mass is physically added to the surface, such as occurs in the various stages of biosensor construction, the frequency of this oscillation must change in order for it to remain in harmony. This change in frequency is monitored and allows for real-time measurements of the changing properties (mass) of the system and, as such, it is a useful technique to monitor impedimetric biosensor assembly (Corry et al., 2003; Rodgers et al., 2010).

Fluorescence is a more traditional, biochemical approach which can be employed to monitor biosensor assembly (Queirós et al., 2013). As a well-established laboratory technique it is relatively easy to fluorescently tag different components of the biosensor assembly. By looking at the addition of different fluorescently tagged components, or conversely quencher moieties, we can monitor the stepwise assembly of the biosensor.

5. Commercialising impedimetric biosensors: from laboratory to field

The demand for reliable and user-friendly medical diagnostic tools that can be used at the point-of-care is well established (Soper et al., 2006; Yager et al., 2008). Point-of-care may refer to use in a clinical environment, field use by a clinician or other operative, or self-use by a patient. Commercial point-of-care diagnostic devices are currently widely available for amperometric and potentiometric electrochemical biosensors, e.g. glucose home blood sugar monitors (amperometric, (Newman and Turner, 2005)), iStat Portable Clinical Analyser (amperometric and potentiometric, see Section 5.3.3). However, at the time of writing, no impedimetric biosensors have made the transition from successful research laboratory trials to mass produced point-of-care diagnostics. In this section, we outline the main considerations associated with the transition of an impedimetric biosensor from research laboratory to widespread use.

5.1 Components of a point-of-care diagnostic device

A point-of-care impedimetric biosensor for clinical applications requires the following components to be considered:

1. Impedimetric biosensor

The diagnostic device may be designed to incorporate one, or multiple, single-use biosensors that are discarded following measurement or that are used repeatedly following a process of regeneration after use (Section 5.4). The incorporation of an impedimetric biosensor into a point-of-care device includes a biorecognition element and a transducer (Section 2.1).

2. Sample collection, delivery and processing capability

Typical fluids to be analysed in medical diagnostics include whole blood and urine (Connolly, 2004), saliva (Arya et al., 2010; Castagnola et al., 2011), and surgical drain fluids (Simmen et al., 1994). A portable, diagnostic point-of-care device will require a method of fluid delivery to the biosensor surface. Some biosensors may also require the sample fluid and/or the sensor itself to be processed prior to sample delivery to the sensor surface, including for example washing and dilution preparation steps (Rivet et al., 2011). In such cases the device will need to deliver and control the flow of non-sample fluids such as reagents or buffers, in addition to the target fluid sample. This is commonly achieved by means of fabricated microfluidic platforms or other flow chamber systems (Section 5.3). The method of sample (e.g. blood, urine, surgical drain fluid) collection may be incorporated into the design of the diagnostic device, such is the case for some models of amperometric blood glucose monitors (Newman and Turner, 2005), or the sample may be collected separately and subsequently delivered to the diagnostic device.

3. Signal analysis capability

A compact battery-operated or rechargeable potentiostat device has to be incorporated into the point-of-care device to enable processing of the electrochemical signal without connection to a base unit (such as a personal computer or laptop), that also provides a built-in display of data in a form which will be meaningful for the user. Stand-alone portable potentiostats are already commercially available, for example the PG581 Potentiostat - Galvanostat model from Uniscan Ltd. (2013b). There will be a requirement, however, for any such commercially available technology to be adapted for use in a point-of-care portable diagnostic device capable of handling fluid samples.

4. Sample disposal

Consideration also needs to be given to how the patient sample will be disposed of post-analysis. One approach may be to encapsulate the biosensor elements and fluid processing capability into single-use cartridges, as used in the commercially available i-STAT Portable Clinical Analyzer (Section 5.3.3). Here, the processed sample is disposed of along with the cartridge. Where the biosensors are incorporated into reusable flow chambers (e.g. microfluidic platforms, Section 5.3.2) however, the chamber will require flushing with a suitable regenerating or cleaning buffer, and the used sample fluid collected externally.

5.2 From research to point-of-care

There are four key factors that influence the successful transition from research device to point-of-care deployment:

5.2.1 Portability and component integration

The portability of a diagnostic biosensor device will determine the range of point-of-care uses for which it can be deployed. For example, the extent of global application as determined by the scale of practical use in developing countries (Yager et al., 2008). Portability is therefore a desirable feature, particularly in maximisation of the commercial opportunities for the device.

A likely outcome of the desire for portability is the miniaturisation of the sensor's transducer components and the fluid delivery platform (see Section 5.2.2). Integration of the sensor into a unit containing sample processing capability and built-in analytical signal processing instrumentation will also be necessary. Therefore, consideration must be given to the transducer surface electrochemical properties following miniaturisation. This includes the positioning of the transducer components to ensure a stable reference potential is maintained during measurements. Sometimes, miniaturisation of components may not always be possible and more flexible sensing elements may be preferred. For example, protein-based nanopores for DNA detection are relatively fixed in size and shape compared to artificial alternatives, and therefore do not translate so well to incorporation into microfluidic platforms (Mir et al., 2009).

The intended end use of the portable measurement device will also be a significant factor in determining the complexity of the required component integration. For example, efficient and realistic cancer diagnostic and prognostic tools typically require measurement of a number of different analytes or biomarkers (Soper et al., 2006). Functionality with respect to multiple target analytes may require the integration of multiple sample preparation steps, as well as the challenge of managing sample delivery to multiple sensor surfaces.

5.2.2 Biosensor performance capability

To some extent, the performance level required of a biosensor-based portable diagnostic device will be determined by its end use but performance will ultimately need to be comparative with accepted laboratory based measurement techniques; the so called "gold standard" methodology. The performance level will typically be assessed in terms of the selectivity of the incorporated biosensor for its target analyte(s) and response to interfering species in the analysis sample, the level of accuracy in comparison to established laboratory techniques, and the reliability and sensitivity of the biosensor (Eggins, 2002); all in relation to the sensor incorporation into the point-of-care device. The reliability of a portable biosensor diagnostic tool is dependent upon the reproducibility of the integrated sensor response over repeated use, within pre-determined levels of accuracy. Additionally, the ability of the integrated biosensor(s) to produce a stable electrochemical signal with each use is a further indicator of the overall reliability.

Another performance factor for consideration is the stability of the biosensor unit in response to long-term storage (Ronkainen et al., 2010), including storage in different conditions of temperature and humidity. The latter may be a particular issue when the biosensor-based diagnostic device is to be used in low resource locations such as some developing countries where access to suitable storage facilities (e.g. refrigeration) may be limited (Yager et al., 2008).

5.2.3 Development costs

Development costs for a point-of-care impedimetric biosensor diagnostic device include the time and effort put into research activity, fabrication costs for the device structures (such as microfluidic platforms), other material costs, and product testing costs. The fabrication costs are determined by the choice of materials and the fabrication methods involved. The cost of the fabrication methods in turn is related to the complexity of the component integration requirements, as discussed in Section 5.2.1.

Ultimately, the development costs need to be considered in relation to the size of the global market for the biosensor diagnostic product. The market size in turn is determined by the diagnostic potential and relevance of the biosensor unit, and the performance issues discussed in Section 5.2.2 (Luong et al., 2008).

Overall, portable impedimetric biosensors are regarded as a potential low cost alternative to existing laboratory based medical diagnostic techniques (Malhotra and Chaubey, 2003). Hence, there are both commercial and practical incentives to seek methods of integration and fabrication that minimise manufacturing costs whilst retaining a high level of performance.

5.2.4 Usability

To ensure widespread common use of a portable diagnostic device it is preferable that the device is easy to use. Certainly, there should be no requirement for the end user to have detailed knowledge in relation to the device operation e.g. biosensor and electrochemical knowledge. This is particularly important where the end users are occasional users, such as members of the public operating the device at home, and patients in developing countries where restricted access to medical facilities and personnel may be a limiting factor for timely diagnosis (McNerney and Daley, 2011; Yager et al., 2008).

5.3 Sample delivery & processing

This sub-section considers the approaches used to deliver a sample to a portable biosensor diagnostic device for the production of a usable measurement signal. Specifically, we consider the guiding principles for sample delivery & processing, outline the use of microfluidic platforms, and identify the specific approaches taken in current commercial products.

5.3.1 Guiding principles

The method of sample collection and delivery to the point-of-care device may vary dependent upon the fluid to be analysed, whilst even for a given fluid type alternative collection and delivery methods may be considered. Some amperometric blood glucose monitors deliver small samples of whole blood (produced by pricking a finger with a sterile single-use needle) onto a disposable test strip containing the biosensor. The test strip is connected temporarily to the signal processing unit for electrochemical measurement (Newman and Turner, 2005). Alternatively, the sample may be collected separately (e.g. urine, blood, surgical drain fluid) and delivered to the portable device using a sterile syringe, as required for operation of the i-STAT Portable Clinical Analyzer (Section 5.3.3) (Erickson and Wilding, 1993).

A major benefit of the miniaturisation process associated with portability is the potential to analyse very small sample sizes, and to minimise the use of any other reagents required for the operation of the biosensor. This can help maintain low operational costs, but also keeps the production of any waste fluids to a minimum for systems that utilise a reusable flow chamber. The design of the sample delivery and processing system therefore needs to facilitate the processing of small fluid volumes e.g. less than one millilitre.

To maintain and maximise portable utilisation, fluid control mechanisms need to be designed to enable integration into the device without compromising portability. Additionally, when incorporating fluidics systems into a measurement device it is important to consider how the design of the flow chamber and the rate of fluid delivery influence the mean concentration of the target analyte within the incoming fluid. This relationship is due to the principles of fluid behaviour within channels and other flow domains, such as the development of flow recirculation at sufficiently high flow rates. This is of particular relevance when the sample to be analysed is introduced into a flow chamber already containing fluid (e.g. reagent or buffer) (Pike et al., 2013).

5.3.2 Mini- and micro-fluidic platforms

A common approach for the application of biosensors is to incorporate the sensor(s) into a mini- or micro-fluidic platform. The term *microfluidics* can be defined as "the science and technology of systems that process or manipulate small (10^{-9} to 10^{-18} litres) amounts of fluids, using channels with dimensions of tens to hundreds of micrometres" (Whitesides, 2006), but hereafter will be used to refer to a type of flow domain independent of scale (i.e. mini- and micro-fluidic systems will not be distinguished unless it is necessary to do so for clarity). The integration of sensor and microfluidics platform is often referred to as a type of lab-on-a-chip (LOC) device or Micro Total Analysis System (μTAS) (Reyes et al., 2002).

The choice of material for a microfluidic platform is selected according to its compatibility with the fluids to be analysed, and with the materials used in the construction of the sensor. The choice of material can also have a significant influence on the cost of the analytical device as a whole.

Popular materials used for platform fabrication include polymers such as polydimethylsiloxane (PDMS), which is often used because of physical properties that provide for high biocompatibility and flexibility in flow domain fabrication designs (Kiilerich-Pedersen and Rozlosnik, 2012). However the chemical properties of PDMS can limit the range of applications, and alternative approaches to fabrication have been developed. These include the use of cyclo-olefin copolymers (COC), thermoplastics such as poly(methyl methacrylate) (PMMA) and polycarbonate (PC), and paper-based platforms. Fabrication using these materials is also more suitable to mass production techniques than PDMS and they therefore have commercial potential due to the relative low production costs (Rivet et al., 2011).

5.3.3 Disposable cartridges

One approach that has been successfully adopted for the integration of microfluidic platform, sensing capability and analytical instrumentation for point-of-care applications is the use of single-use disposable cartridges.

The i-STAT Portable Clinical Analyzer is a commercially available example of this approach (2013a; Erickson and Wilding, 1993). This system accommodates a number of different cartridges, each of which is used to analyse small blood samples for multiple or single target diagnostic markers. The diagnostic markers analysed include electrolytes and chemical indicators (e.g. sodium, potassium, chloride, and urea nitrogen), haematological indicators (haematocrit and haemoglobin) and blood gases. The cartridges contain a calibrant solution and any required reagents, a microfluidic platform and fluid control system, and a series of biosensors constructed upon a silicon chip substrate. The cartridges are inserted into a handheld analytical device for operation. The measurement method varies depending on the target diagnostic markers, but includes potentiometry and amperometry.

5.4   Sensor regeneration

As discussed, one key barrier to the widespread commercialisation and usage of impedimetric biosensors is currently their cost (Luong et al., 2008). This is due to the electrode components as well as the analysis hardware used which is currently restricted to large, research level systems (Hall, 2004). In enabling biosensor regeneration, that is the ability to re-use the sensor chips multiple times, the cost per test can be reduced, thereby encouraging their wider adoption in clinical settings. Work has been done to enable re-useable sensors but, although the older amperometric sensors have been demonstrated to be extensively reusable (Manso et al., 2008; Vidal et al., 2004), more modern types of biosensors, including impedance-based platforms, have yet to be capable of regeneration to the same extent.

For some analytes, such as metabolites and persistent pathogens or biomarkers of chronic disease states, continuous or quasi-continuous monitoring is desirable. Although single-use sensors have been demonstrated for these applications, one problem is chip-to-chip variance. This can be minimised by controlling biosensor construction, but a reusable sensor would be preferred. This would mean that chip-to-chip variability would be eliminated and, although there may be some signal loss with repeated use, this can be accounted for much more easily.

There is currently intense research into the possibility of implantable biosensors for continuous monitoring (Kotanen et al., 2012). Continuous monitoring provides more detailed information, can detect brief fluctuations in analyte levels, and also removes the need for periodic sampling, such as taking blood, representing an easier option for a patient. One issue, however, with implantable sensors is the product lifetime. A common problem is the signal loss over time. Therefore, it is paramount to ensure that any implantable biosensor has good regeneration characteristics. This problem is not unique to implantable sensors as offline, point-of-care biosensors may also require regeneration. As sample interrogation is performed offline, in these types of sensor, the risk of patient-to-patient contamination is removed.

The regeneration of biosensors is enabled by selectively reversing the interaction between the target analyte and the bioreceptor (Figure 5-1). For example, antibody:antigen interactions are well defined and their binding biophysics are understood comprehensively (Conroy et al., 2009). The forces between antibody and antigen comprise a variety of non-covalent interactions which can be reversed by altering the buffer conditions, thus allowing

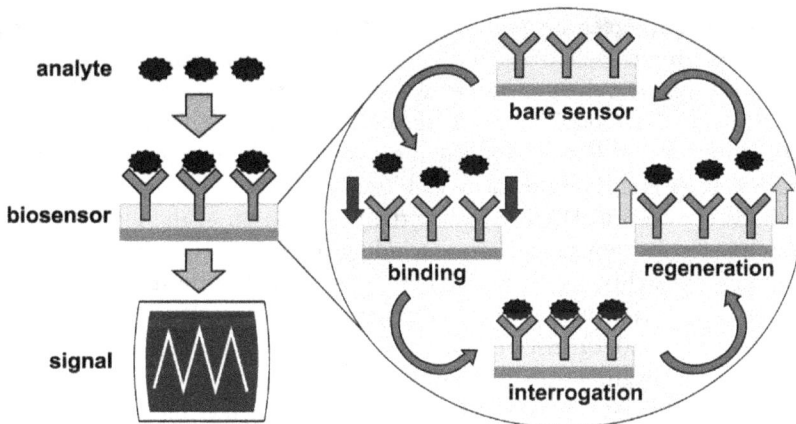

Figure 5-1 Schematic showing biosensor regeneration.

regeneration in different antibody-based biosensor systems (Andersson et al., 1999; Bright et al., 1990).

Work has also been done on the regeneration of DNA aptamer based impedimetric sensors (Queirós et al., 2013). One advantage of using oligo-nucleotide binders is their very regular, well defined chemical properties. The sugar-phosphate backbone on which they are based has been widely studied and the underlying binding physics are often predictable. Nucleotides can be selectively denatured by raising the temperature and, as forces between the analyte and oligonucleotide tend to be governed by charge, they can also be screened relatively easily by changing pH or ionic strength. Due to the resilience of the nucleotide residues and backbone, DNA aptasensors can be regenerated reasonably well (Lazerges et al., 2006; Queirós et al., 2013; Tedeschi et al., 2005).

In order to achieve regeneration, a low pH glycine buffer has been employed in many different biosensor systems (Dillon et al., 2003; Dillon et al., 2005; Drake and Klakamp, 2011; Hong et al., 2009; Huang et al., 2010; Kandimalla et al., 2004; Wijesuriya et al., 1994; Xu et al., 2013). The buffer-ing range of glycine, a natural amino acid, is between pH 2.2-3.6. The glycine molecules help to solvate the analyte away from the receptor in a process which is driven entropically. In some reports, the acidic glycine is included with a cocktail of other detergents, salts and chaotropic agents, all of which encourage dissociation of the analyte from the bioreceptor (Andersson et al., 1999). Similarly, high pH buffer systems, for instance sodium hydrox-ide, have been shown to reverse binding of the receptor:analyte complex (Albrecht et al., 2008; Bryan et al., 2013; Chen et al., 2010; Mattos et al., 2012; Michalzik et al., 2005; Steegborn and Skládal, 1997; Xu et al., 2013). The change in pH causes a change in the ionic strength of the solution; this has the effect of screening many charge interactions between the analyte and the bioreceptor as well as, certainly in protein models, transiently denaturing the bioreceptor, thereby allowing the analyte to dissociate more freely.

Although the most success in terms of sensor regeneration has been achieved in optical biosensors, the interfacing of the bioreceptor with the sensor surface is similar across many biosensor types. This means that these methods can be translated to enable the regeneration of the more modern types of electrochemical sensors, including impedimetric sensors. Present work confirms this, albeit in the early stages, with a limited number of regeneration cycles proven for electrochemical sensors (Bryan et al., 2013; Huang et al., 2010; Queirós et al., 2013; Xu et al., 2013; Yun et al.).

Biosensor architecture is a key consideration when regenerating an impedimetric affinity biosensor as the regeneration buffer must not alter the signal of the sensor itself. This may help to explain why electrochemical sensors have achieved limited success as factors such as pH and ionic charge may interfere with the biosensor signal. Therefore, it must be ensured that the regeneration conditions do not interfere with the substrate, be it SAM or polymer. For use in a medical point-of-care device, it is likely that regeneration would be achieved using a microfluidic set-up with a regeneration cycle to improve ease of use and control regeneration parameters, as discussed in the previous section.

5.5 Barriers to commercialisation

The potential of impedimetric biosensors as low cost diagnostic tools is well understood and has been presented throughout this monograph. However, this potential currently remains based in the research laboratory as the breakthrough to integration into a commercial mass produced point-of-care device has yet to be made.

The main current barrier to the transition of impedimetric biosensors from the research lab to incorporation into point-of-care diagnostic devices is the stability and reliability of impedimetric biosensors compared to other sensor types (Prodromidis, 2010). Reliability in this sense refers to both the ability of the biosensor to produce a demonstrable accurate measurement signal with repeated use (where applicable), and from one biosensor unit to another. This leads to a significant problem in terms of scaling-up biosensor production to the mass production level.

Also, biosensor systems that require the integration of multiple preparation and processing stages will lead to complex microfluidic platform fabrication and fluid handling capability. This could prove to be both a challenging technological hurdle, and relatively high on fabrication costs (Choi et al., 2011; Mir et al., 2009). However, production costs are always linked closely to volume of production, and fabrication on a large scale, e.g. as with glucose sensors, brings about a dramatic reduction in unit cost.

6. Case studies of impedimetric biosensors for medical applications

Many research reports have been published in recent years detailing the impedimetric detection of various types of medically-relevant analytes and/or biomarkers, including small molecules such as metabolites and drugs, proteins and peptides, nucleic acids, and whole microorganisms such as pathogenic viruses and bacteria. Traditional, detection methods within clinical biochemistry are composed of several steps; sample collection, sample preparation or labelling, multi-step sample incubation and long processing times, sometimes up to weeks. Many of these methods, such as polymerase chain reaction (PCR), require expensive equipment and skilled laboratory technicians. Biosensors offer a much faster, inexpensive diagnosis that can be obtained in a matter of minutes at the point-of-care. As discussed previously, impedimetric biosensors offer the key advantages of small device size, low sample volumes, usually label-free detection with minimum or no sample preparation steps, at low cost.

A plethora of published reports have described the laboratory-based detection of a range of different medical analytes by impedimetric biosensing (Table 6-1). In this section, several detailed examples of impedimetric biosensors for the detection of medically-relevant analytes will be presented. As discussed in Section 4, the nature of the analyte can determine various parameters within the biosensor system, including the biorecognition element and the transducer surface, which in turn can influence the choice of electrode material. For this reason, the case studies presented here are categorised according to the type of analytes.

However, it should be noted that an alternative method of categorising impedimetric biosensors could be based on sensor functionality; some sensors are used to provide a rapid diagnosis of a life-threatening disease state, for instance confirming a heart attack, whereas other sensors are used in situations where an immediate diagnosis is not critical or for continuous monitoring.

6.1 Whole cells and pathogenic microorganisms

Pathogenic bacteria and viruses cause substantial mortality and morbidity worldwide. Both are associated with an enormous number of human diseases, many of which are fatal if not detected in time. For example, according to data from the World Health Organisation, HIV infection was responsible for 1.7 million deaths worldwide in 2011; in the same year, 8.7 million people were infected with tuberculosis with 1.4 million deaths reported. Thus, the early detection of pathogens can save lives and also saves money in terms of negating costly treatments and intensive care stays. Traditional bacterial and viral detection methods include cell culture and PCR, which require time, skilled personnel, complex lab equipment and post-detection

Table 6-1 Published examples of impedimetric biosensors for clinical applications.

Analyte	Electrode	Transducer surface and tethering	Bioreceptor	LOD	Reference
		Whole cells			
E. coli	Interdigitated gold microelectrode	Physisorption	Ab and T4 phage	10^4–10^7 cfu/ml	(Mejri et al., 2010)
E. coli O157:H7	Gold millielectrode	Aniline-glutaraldehyde	Ab	10^2–10^7 cfu/ml	(Chowdhury et al., 2012)
E. coli O157:H7	Anodised aluminium oxide membrane	Silane-HA	Ab	10 cfu/ml	(Joung et al., 2013)
Cancer cells	Gold electrode	Activated absorption	Con A	234 cells/ml	(Hu et al., 2013)
Endothelial progenitor cells	Gold electrode array	SAM/EDC/NHS	Ab	720 cells in 12 μl	(Ng et al., 2010)
		Viruses			
Dengue virus	Porous alumina membrane	Physisorption	Ab	NA	(Peh and Li, 2013)
Adenovirus	Screen printed gold electrode	Sulfo-SMCC	Half-Ab	10^3 virus/ml	(Caygill et al., 2012)
		Proteins and peptides			
Tumor biomarker CA125	Gold nanoparticle on screen printed graphite electrode	mSAM-EDC/NHS	Ab	6.7 U/ml	(Ravalli et al., 2013)
Aflatoxin m1	Gold electrode	SAM-gold nanoparticle	Thiol-modified ssDNA	1–14 ng/ml	(Dinçkaya et al., 2011)
Deep venous thrombosis biomarker	Gold microelectrode	Single walled carbon nanotube	Fab	0.1 pg/ml	(Bourigua et al., 2010)

Proteins and peptides

Analyte	Substrate/electrode	Immobilization	Receptor	Detection limit	Reference
Human serum albumin	Glass and gold	3-aminopropyltriethoxysilane (APTES)	Ab	2×10^{-4} mg/ml	(Chuang et al., 2011)
Cancer biomarker Murine double minute 2 (MDM2)	Gold electrode	SAM-1,4-phenylene diiso-thiocyanate (PDITC)	Ab	0.29 pg/ml (PBS); 1.3 pg/ml (mouse brain)	(Elshafey et al., 2013)
IL-6 antigen	Silicon wafer	SAM/EDC/NHS	Ab	0.01–100 fg/ml	(Yang et al., 2013c)
VEGF	Gold electrode	SAM/EDC/NHS	VEGF receptor	10–70 pg/ml	(Sezgintürk, 2011)

DNA/RNA

Analyte	Substrate/electrode	Immobilization	Receptor	Detection limit	Reference
rRNA of *Mycobacterium tuberculosis*	ITO coated glass	Adsorption onto zirconium graft	ssDNA	NA	(Das et al., 2011)
DNA of fungus *P. sclerotigenum*	Gold-magnetic nano particle	poly(allylamine hydrochloride) SAM	ssDNA	NA	(Silva et al., 2013)

Small molecules

Analyte	Substrate/electrode	Immobilization	Receptor	Detection limit	Reference
Digoxin	Microarray carbon electrode	Biotin-avidin	Ab	0.1 ng/ml	(Barton et al., 2009)
Oestrogens	Gold electrode	SAM/EDC/NHS	Receptor	1×10^{-13}M	(Kim et al., 2012)

Abbreviations: Ab, antibody; EDC, ethyl((dimethylaminopropyl) carbodiimide; Fab, fragment antigen binding region; HA, hyaluronic acid; ITO, indium tin oxide; mSAM, mixed self-assembled monolayer; NHS, N-hydroxysuccinimide; SAM, self assembled monolayers; ssDNA, single stranded DNA; sulfo-SMCC, 4-(N-maleimidomethyl) cyclohexane-1-carboxylic acid 3-sulfo-N-hydroxysuccinimide ester sodium salt.

processing. Moreover, PCR in particular is prone to generating false-positives. To catch bacterial and viral infections in time and to facilitate effective medical care, the detection time should be minimal and biosensing devices must be user-friendly with point-of-care facility.

Impedimetric biosensing can facilitate the early detection and quantitation of viruses or bacteria in order to confirm a particular infection and also to determine the stage of infection based on the infectious load. This can be performed on patient samples without the need for labelling reagents and can be used for continuous patient monitoring, e.g. an impedimetric sensor comprising a flow cell can be used for in-line testing of drain fluid, blood and so on. Like all other impedance methods, the choice of transducer, linking chemistry and selection of bioreceptors heavily influence the sensitivity of the results. Several examples of recent research targeting whole cells and viruses are presented here, with a consideration of key aspects of biosensor construction and the resulting success of the platform.

6.1.1 Whole bacterial cells

Impedimetric biosensors have been used to detect a wide range of bacteria, including *Escherichia coli* (*E. coli*) (Escamilla-Gómez et al., 2009), *Salmonella typhimurium* (Pournaras et al., 2008), *Staphylococcus aureus* (Tan et al., 2011) etc. The detection and quantification of whole bacterial cells can give direct information on the infectious load present. *E. coli* is the widely used model bacterium in impedance-based biosensor research because of its ease of culture and manipulation in the laboratory environment. Most strains are non-pathogenic, although certain ones, such as O157:H7, can cause intestinal infections and can be fatal in elderly or fragile patients.

Published *E. coli* impedance sensors can be categorised according to various properties, including the type of electrode material, choice of bioreceptors, method of bio-conjugation, type of sample analysed etc. The detection of *E. coli* in buffers, serum, food samples etc. has been explored extensively and these findings have been translated to the detection of medically-relevant, pathogenic strains within patient samples. A variety of different biosensor architectures have been used to construct *E. coli* sensors; these include different base layers, including SAMs (Geng et al., 2008), polymers (Chowdhury et al., 2012) and nanoporous membranes (Joung et al., 2013), and a wide variety of bioreceptors, such as antibodies (Barreiros dos Santos et al., 2013), lectins (Gamella et al., 2009) and bacteriophage (Mejri et al., 2010; Shabani et al., 2008). Some recent, innovative findings are discussed in more detail in the next few paragraphs.

The early detection of pathogens at low concentration is vital for fast, effective treatment. Recently, the detection of a very low number of *E. coli* was reported in a study where a SAM-based impedimetric immunosensor was fabricated upon gold disk electrodes (Barreiros dos Santos et al., 2013). The study confirmed the detection of only two colony-forming units per millilitre (c.f.u/ml) of *E. coli* in phosphate buffered saline, although its response in more biologically-relevant sample medium was not reported. SPR was used to optimise layer-by-layer sensor construction (Section 4.2.4), and a fluorescence-based indirect assay was used to confirm analyte binding. *Salmonella typhimurium* was used as a control organism and less than 20% non-specific binding was reported for the sensor.

Polyaniline was used as base layer to construct an antibody-based immunosensor against *E. coli*. Whole antibodies were immobilized onto the polyaniline surface by covalent cross-linking. This sensor showed high specificity and was able to discriminate other strains when tested (Chowdhury et al., 2012).

Bacteriophage, or simply phage, are viruses which can infect certain species of bacteria with high specificity and which replicate inside their host. This specific interaction can be exploited to design biosensors where phage are immobilised as bioreceptors onto the sensor surface where they can bind their specific bacterial targets. A study based on this phenomenon used T4-phage, a phage specific to *E. coli*, as a bioreceptor on gold electrodes to capture *E. coli* (Mejri et al., 2010). EIS was used to characterise the binding of bacteria to phage on the sensor surface. Increasing EIS signal with increasing bacterial concentration confirmed the specific binding of bacteria and then, over time, a decrease in EIS confirmed the phage-mediated lysis of bacteria on sensor surface. This additional lysis-induced EIS signal helped to confirm that bacterial binding was due to specific attraction and not due to non-specific binding. This report also indicated that phage-based sensors can be more sensitive and specific as compared to antibody-based sensors.

Bio-imprinting is a method where particular biomolecules or cells are deposited on a surface and then washed off, leaving their imprint on the surface. In one recent study, sulfate reducing bacteria (SRB) mediated bio-imprinted films were developed to capture the same bacteria in solution (Qi et al., 2013). First, layers of reduced graphene sheets and chitosan were electrodeposited on indium tin oxide (ITO)-based electrodes, followed by absorption of SRB with a thin coating layer of non-conducting chitosan around the bacteria. Following this, the SRB was washed off the surface to get the bio-imprint on top of the sensor. This imprint was used to capture and quantify target SRB in a range of 10^4 to 10^8 c.f.u/ml using EIS. The sensor was able to distinguish other control strains based on differences in size and shape but the authors recommend its use in combination with additional, specific bioreceptors, such as antibodies, for improved performance.

6.1.2 Human cells

There are other whole cell types besides bacteria, for instance cancer cells, which also represent a good target for impedance-based biosensing. In 2008, the World Health Organisation reported an estimated 7.6 million deaths were due to cancer. Therefore, the direct detection of cancer cells in a mixed sample with healthy cells can be a useful diagnostic tool. Concanavalin A (Con A), a type of lectin, was immobilised onto a gold surface for the specific and selective detection of cancer cells. Lectins are sugar binding proteins which can specifically bind to altered oligosaccharides that are expressed uniquely on the surface of certain cancer cells. The underlying principle was the ability of Con A to bind altered carbohydrates in cancer cells (BEL-7404), a human hepatocellular carcinoma expressing membrane associated glycoprotein gp43. The sensor showed good selectivity over a normal liver cell line (L02) when used as control. The sensor response was very fast and accurate compared to traditional methods of cancer marker detection (Hu et al., 2013).

Cardiovascular disease accounts for an estimated 17.3 million deaths worldwide with many more people at high risk. Whole cell biomarkers of cardiovascular disease can also be investigated using impedance biosensors. Endothelial progenitor cells (EPCs), an important cardiovascular marker, were detected successfully using a microfluidic impedance sensor. Using dielectrophoresis, a technique which requires very small sample volumes, just 720 EPCs were detected in a 12 µl sample of peripheral blood mononuclear cells (Ng et al., 2010).

6.1.3 Viruses

The detection of whole viruses can also be employed to identify and quantify the number of infectious particles in any biological sample. To identify and distinguish particular strains from others can also provide useful information in certain situations. Impedimetric methods thus provide a quick and sensitive solution in detecting whole viruses in medical samples, avoiding complicated and costly molecular biological identification techniques.

The influenza virus causes around 500,000 deaths worldwide each year with many patients requiring hospital treatment. A bio-mimetic surface comprising an mSAM of octanethiol and octagalactoside linked to sialic acid was constructed on a gold surface to mimic the surface of host cells to which the virus binds (Wicklein et al., 2013). Two different types of surface sialic acid were used as bioreceptors specific to human (H1N1) and avian (H5N9) strains, respectively, and these were able to detect and distinguish these strains successfully by monitoring the change in impedance upon viral binding. The sensor was rapid, accurate and very sensitive and could potentially be used to screen strains of influenza virus in patient samples. In another study, an mSAM-based immunosensor on a gold surface was fabricated to detect influenza A virus H3N2 (Hassen et al., 2011) with a limit of detection of 8 ng/ml of virus. The immunosensor was constructed using a biotin-Neutravidin bridge to tether biotinylated anti-H3N2 antibody onto the mSAM surface. The sensor was investigated in a bio-mimic solution containing a mixture of hepatitis B virus and serum in phosphate-buffered saline to show that a distinguishable impedance signal can be obtained in this complex bio-mixture.

Dengue fever is mosquito borne viral infection causing fever and leading to a fatal condition known as severe dengue. Because the early symptoms are very similar to other, less serious fever-related conditions, its delayed identification or initial misdiagnosis often leads to development of the more severe disease form. According to the World Health Organisation, half of the global population is at risk to this infection and its early detection can reduce the fatality rate to below 1%. In a recent study, alumina membrane was used as the base electrode material to detect dengue virus (Peh and Li, 2013). Anti-dengue antibody was immobilised in the membrane pores; analyte recognition changed the pore resistance which could be correlated with virus binding. This sensor also showed specificity over other dengue strains and was able to detect the virus in spiked serum samples. As an alternative to whole dengue virus detection, infection-related glycoprotein sensing by specific lectins (Oliveira et al., 2011) and non structural protein NS1 detection by anti-NS1 monoclonal antibody (Cavalcanti et al., 2012) have been reported.

Caygill et al recently described a copolymer-based immunosensor to detect adenovirus, which is a life threatening virus for immunocompromised patients (Figure 6-1) (Caygill et al., 2012). In this study, aniline and its derivative 2-aminobenzylamine were copolymerised onto commercially available screen-printed gold electrodes. Specifically-oriented chemically reduced and purified half-antibody fragments, as described in Section 4.2.2, were immobilised on the transducer surface. The sensor was specific and sensitive with a limit of detection of 10^3 virus particles per ml of sample.

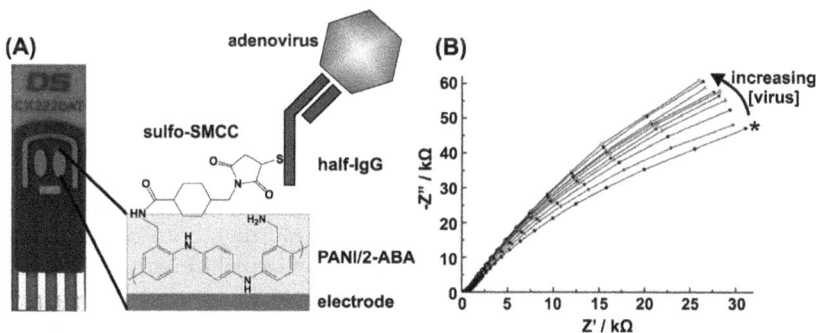

Figure 6-1 Impedimetric detection of whole adenovirus using half-antibodies. (A) Schematic of biosensor construction. Half-antibodies (IgG) against whole adenovirus particles were tethered via sulfo-SMCC to screen-printed gold electrodes coated in the co-polymer PANI/2-ABA (comprising polyaniline and 2-aminobenzylamine). (B) Nyquist plot shows that increasing concentration of adenovirus caused an increase in impedance (* = no analyte). Data modified from (Caygill et al., 2012).

6.1.4 Other markers of pathogenic infection

Many impedance sensors have been used to detect alternative markers of pathogenic infection, including cell surface proteins, cell fractions, extracellular secretions, toxins, DNA and RNA. The detection of such products is sometimes more sensitive than detecting whole cells to confirm the presence of a particular organism, but also is often burdened with extra sample processing steps.

Tuberculosis (TB), caused by *Mycobacterium tuberculosis*, is one of the most lethal pathogenic strains for human infection. Almost one third of the world's population is infected with the deadly bacteria. Normally, a chest x-ray is performed to identify the infection, which is then confirmed by blood test. This is a lengthy procedure which requires specialist medical personnel. Recent advances in nanotechnology have assisted in the development of biosensors for the faster, simpler detection of TB (Wang et al., 2013a). Data from acoustic-based impedance sensors (He et al., 2003), magneto-elastic sensors (Pang et al., 2008) and quartz crystal microbalance (QCM) (Hiatt and Cliffel, 2012) have been reported to detect mycobacteria, but published data on the detection of whole cells using EIS are limited. This may be due to the highly contagious nature of the bacteria and the risks associated with laboratory-based biosensor testing. However, the detection of DNA from mycobacteria using EIS has been reported (Das et al., 2011). In this study, zirconia-grafted carbon nanotubes were deposited on ITO, followed by linking single-stranded DNA that binds specifically to bacterial ribosomal RNA (rRNA) as bioreceptors. The EIS signal increased in linear fashion with increased concentration of bacterial rRNA. However, a drawback of this indirect bacterial sensing method is the multistep isolation of bacterial DNA/RNA, which complicates its point-of-care use in a home or hospital environment.

Human immune deficiency virus (HIV), the causative agent of acquired immunodeficiency syndrome (AIDS), is another enormously important medical analyte. The current detection methods are enzyme-linked immunosorbent assays (ELISA) and western blotting, both of which are laboratory-based techniques. There is commercial colorimetric HIV test (OraQuick™) for oral swabs, serum or blood, which detects the presence of virus-specific antibodies, which take several weeks to appear in patient sample after infection (http://www.oraquick.com/). To detect infection earlier, for instance by detecting whole HIV viruses, impedance-based sensors are a promising alternative. A very interesting approach was employed to detect HIV-1 protease, a major indicator of HIV-1 infection which is essential for viral assembly, using EIS (Esseghaier et al., 2013). In this study, magnetic beads were linked to the gold sensor surface *via* protease-specific peptides. When the viral protease cleaves the peptide, the beads dissociate and a decrease in EIS signal is observed. The sensor was able to detect as low as 10 pM of viral protease in phosphate-buffered saline within 20 minutes. However, an HIV-sensing method that functions in patient samples has yet to be described.

6.2 Protein and peptide biomarkers of disease

Proteins and peptides serve as biomarkers of a wide range of diseases besides pathogenic infection and their identification can be achieved using impedimetric biosensors.

In addition to the whole-cell approach to detecting cancer by impedimetric biosensing, various protein and peptide markers for cancer have been explored. The breast cancer biomarker VEGF, a peptide growth factor which is over-expressed in certain cancer conditions, has been detected using EIS (Sezgintürk, 2011). Here, the cell-surface receptor protein for VEGF, the VEGF-1 receptor, was linked to a SAM-functionalised transducer surface to construct the biorecognition element. The receptor was able to bind specifically the target analyte over the range of 10–70 pg/ml with high affinity and sensitivity. The authors also confirmed the stability and reproducibility of the sensor performance, which is a basic requirement for a sensor to be commercialised.

As discussed in Section 4.1.2, the nano-modification of electrode surfaces using materials such as carbon nanotubes and gold nanoparticles can help to boost the impedance signal generated by analyte binding by increasing the surface area for bioreceptor attachment. Nano-modified electrode surfaces have proven successful in the detection of cancer biomarkers. Alpha-foetoprotein (AFP), a glycosylated, 70 kDa protein biomarker for hepatocellular carcinoma, was detected using a specific lectin (wheat germ agglutinin; WGA) as the biorecognition element. As described in Section 4.1.2, screen-printed carbon electrodes were coated with single walled carbon nanotubes for lectin attachment. Different concentrations of AFP were analysed using EIS with a detection limit of 0.1 ng/ml (Yang et al., 2013a). The suitability of five different types of lectins as receptors for AFP was investigated using EIS, where WGA was the most suitable, indicating that lectin-carbohydrate profiling against different types of cancer cells could be facilitated by impedimetric biosensing.

The electrodeposition of gold nanoparticles from tetrachloroauric acid solutions upon screen-printed graphite electrode was employed in the detection of the ovarian tumour marker CA125, a 250 kDa glycoprotein, both in phosphate-buffered saline and in serum-containing samples (Ravalli et al., 2013). Faradic EIS was used to characterise each step of sensor construction and in the detection process. The sensor showed lower detection limit of 6.7 U/ml with a positive, linear correlation between change in charge-transfer resistance (R_{ct}) and increasing concentration of CA125.

A protein biomarker for deep vein thrombosis (DVT), a potentially life-threatening condition caused by venous blood clotting, was detected using an impedance-based immunosensor. The biomarker known as D-dimer, a 190 kDa protein, was detected successfully with a limit of detection of 0.1 pg/ml. The biosensor also functioned in rat blood samples containing D-dimer with only 10% variation in R_{ct} compared with the signal generated in synthetic buffer (Bourigua et al., 2010). This shows promise for the functionality of the sensor in patient samples.

As already discussed in Section 6.1.2, cardiovascular biomarkers represent an important potential target for impedimetric biosensing. Myoglobin, a protein marker for acute myocardial infarction (i.e. heart attack), was detected impedimetrically using two different approaches. Lithographically-deposited atomically flat gold electrodes were used as electrode material; as mentioned in Section 4.2, electrode flatness can assist in the deposition and molecular continuity of SAMs. In the first approach, biotinylated anti-myoglobin antibodies were immobilised onto long-chain mSAMs. The biotinylated antibodies were linked to the mSAM surface, also containing biotin groups, *via* NeutrAvidin. The sensor gave a linear response in R_{ct} change with increasing concentration of myoglobin in phosphate-buffered saline and also in the presence of serum, although the serum-spiked samples showed decreased signal compare to buffer-only samples (Billah et al., 2008). In the second approach, thin-layer SAMs generated from 4-aminothiophenol (4-ATP) were deposited onto gold. Half-antibodies were linked to the surface amine groups of the SAM using sulfo-SMCC. The half-antibody based sensor produced a better signal compared with the full antibody system due to the increased bioreceptor packing close to the sensor surface and the specific orientation of the bioreceptors, i.e. the binding sites all faced outwards, thereby reducing noise and signal loss (Billah et al., 2010).

6.3 Small molecules

Small molecules are usually defined as low molecular weight organic compounds of less than 1 kDa in molecular weight. These can include cell signalling molecules, hormones, antibiotics, drugs, toxins and primary or secondary metabolites with medical significance. Impedance is a particularly useful technique for the detection of trace amounts of small molecules, which can be difficult, time-consuming expensive to detect using traditional methods such as mass spectrometry and high-performance liquid chromatography (HPLC).

Oestrogens, a group of primary female sex hormones, are important indicators of menstruation and oestrous cycle. Some oestrogens, such as 17-β-oestradiol, also have carcinogenic properties, making these compounds relevant biosensor targets. Laboratory-based techniques such as HPLC, SPR and ELISA can be used to measure oestrogen levels, but faster, cheaper, label-free detection would be much more useful in a clinical setting. In one study, an impedimetric biosensor was able to detect concentrations of 17-β-oestradiol as low as 1×10^{-13} M using its cellular receptor, oestrogen receptor alpha (ER-α), as the biorecognition element (Kim et al., 2012). Interestingly, the receptor-based sensor showed better performance compared with a similar sensor which employed antibodies against the same analyte. The authors also confirmed the stability of their sensor for up to three weeks in phosphate-buffered saline with a maximum activity loss of twelve percent.

Another steroid hormone, cortisol, has received medical attention because of its reported role in stress-related disorders including Post-Traumatic Stress Disorder (PTSD). An impedimetric immunosensor constructed upon polyaniline-coated nanoparticles which were deposited upon gold electrodes was able to detect cortisol over a range of 1 pM to 100 nM (Arya et al., 2011).

Toxins can also be detected using impedimetric sensors. Aflatoxins, naturally produced by fungi when food products are stored in damp conditions, can be ingested by farm animals or humans and converted to a secondary metabolite, aflatoxin M1, which has long term carcinogenic effects. An aflatoxin M1 biosensor was constructed by using single-stranded DNA specific to aflatoxin M1 as bioreceptor, linked to gold nanoparticles deposited on SAM-coated gold electrodes. The sensor was able to detect aflatoxin in analyte-spiked milk samples over a concentration range of 1–14 ng/ml with a linear relationship between EIS signal and analyte concentration (Dinçkaya et al., 2011).

The detection and monitoring of trace amount of drugs or antibiotics can be achieved with impedimetric sensing systems. Chloramphenicol, an antimicrobial drug, needs to be monitored in patients because, like other drugs, an overdose can cause adverse effects including renal failure. In one study, ultrasensitive detection (1×10^{-16} M) was achieved using a multilayer sensor where anti-chloramphenicol antibodies were attached to nanoparticles (Chullasat et al., 2011). Microfluidic sample injection methods were employed to record chrono-impedance at an optimised frequency. The sensor was shown to be selective over other structurally-related small molecules such as florfenicol, thiamphenicol and chloramphenicol base where no significant signal was observed. The sensor was also successfully regenerated for up to 45 analysis cycles. As discussed in Section 5.4, regeneration involves the detachment of bound antigen, thereby cutting costs by allowing the sensor to be used repeatedly.

In conclusion, it is clear that the design and fabrication of impedimetric biosensors requires optimisation and personalisation depending on key factors such as the nature and size of the target analyte, as well as the type of sample in which the analyte will be found. Various methods of sensor optimisation, particular nano-scale modifications to the electrode surface, can achieve better signal and higher specificity. Whilst robotics can minimise the batch-to-batch variability of sensors, this also has the downfall of drastically increased research costs. As discussed in Section 5, factors such as the ease of miniaturisation and low cost of the system will facilitate the commercialisation of impedimetric sensors.

7. Conclusions and future perspectives

Medical diagnostics is an industry that is expanding rapidly, with the global biosensor market expected to exceed $US18 billion by 2018. Electrochemical biosensors, as exemplified by the blood glucose monitor, have the potential to revolutionise medical diagnostics and patient monitoring. They offer cheap, automated and easy-to-handle platforms which combine sample processing and analysis to give a read-out in minutes. Impedimetric biosensors are enormously attractive medically, as they offer a rapid, portable, inexpensive and highly sensitive test that can be performed at the point-of-care by a non-specialist user.

However, the widespread commercialisation of electrochemical biosensing devices for medical applications has lagged far behind the recent advances made in laboratory-based sensor systems. This is particularly applicable to the relatively young field of impedimetric biosensing which has only developed over the last 30 years. As we have discussed, a plethora of research publications have reported over the last decade the generation of highly sensitive, robust platforms for the impedimetric detection of diverse analytes, ranging from small molecules up to whole bacteria and viruses (Millner et al., 2012). However, the translation of these laboratory-based technologies to small, hand-held sensors that are available in the clinic has proven difficult.

Although certain commercially available impedimetric sensing methods exist (e.g. the rapid automated bacterial impedance technique (RABIT) for the detection of bacteria (Don Whitley Scientific Ltd., Shipley, UK)), these are large, heavy pieces of equipment that require sample processing, computing power and mains electricity (Yang and Bashir, 2008). A portable impedimetric biosensing platform that can be used at the point-of-care is still lacking. Current research in impedimetric biosensing is working to overcome the current barriers to the commercialisation of point-of-care impedimetric biosensors, which include high costs and lack of specificity in biological samples. Improved methods of electrode fabrication and screen printing, along with nano-modifications of electrode surfaces in order to boost signal whilst minimising non-specific binding, are paving the way forward to develop cheaper, more sensitive and more robust devices. The generation of novel, synthetic bioreceptors and the optimisation of sensor surface chemistry and nano-topology is improving device performance in biological samples such as blood and urine. Advances in microfluidics technology will also assist in sample delivery and analysis. Finally, fostering collaborations between academic and industrial partners will play a critical role in bringing impedimetric sensors out of the laboratory to the point-of-care.

Author biographies

Jo V. Rushworth
Jo received her BSc in Biochemistry in 2005 and her Wellcome Trust-funded PhD in 2012 from the University of Leeds. Jo also carried out research projects at the Sainsbury Laboratory (Norwich, UK) and the Université de Paris-Sud (France). In between her degree and PhD, Jo was a high-school Chemistry teacher. Jo studied the molecular and structural biology of amyloid-beta oligomers, a causative agent of Alzheimer's disease, during her PhD. She is now integrating her Alzheimer's research background with her interest in electrochemical biosensors to develop impedimetric sensors for the specific detection of biologically-relevant amyloid-beta oligomers.

Natalie A. Hirst
Natalie gained a BSc in Experimental Pathology in 2005, before a Medical degree in 2006 from the University of London, Barts and the London School of Medicine and Dentistry. She is a member of the Royal College of Surgeons of England, having passed the requisite examination. She is currently in her final year of study for a PhD with the Bionanotechnology Group at the University of Leeds, having taken time out of full time surgical training. Her current research is the development of electrochemical biosensors for early detection of complications after bowel surgery.

Jack A. Goode
Jack received his BSc in Nanotechnology in 2010 from the University of Leeds, during which time he completed a research project looking at novel photo-activated dental materials. Jack is currently undertaking a CASE-funded PhD at Leeds, in which he is developing multi-array electrochemical biosensors for research and medical diagnostics, in collaboration with an industrial partner. Jack is particularly interested in the modification of antibodies in order to improve their performance as bioreceptors in impedimetric biosensing, as well as biosensor regeneration and the integration of sensors into multi-array systems.

Douglas J. Pike
Doug received his BSc in Geosciences from the Open University in 2008 and then obtained an MSc in Hydrogeology from the University of Leeds in 2010. Doug is currently undertaking his PhD, which combines engineering and electrochemical biosensing, at Leeds. Doug is currently developing an automated, remotely-controlled system for the real-time monitoring of radioactive contaminants in groundwater. Doug's research has already generated a flow cell in which electrochemical biosensors for this application can operate. His current research is focussed upon the integration of environmental biosensors into the novel flow cell.

Asif Ahmed
Asif received his BSc in Biotechnology and Genetic Engineering from Khulna University, Bangladesh, in 2002. After spending two years as a Lecturer at Khulna University, Asif obtained his fully funded MSc in Biomolecular Science from the University of Science and Technology, Korea in 2009. During his MSc he utilised *in silico* pharmacophore modelling and virtual screening to generate novel anti-obesity drug scaffolds to target the serotonin receptor (5-HT2c). Asif currently holds the post of Assistant Professor at Khulna University whilst he is carrying out his PhD at the University of Leeds. Asif's current research focuses upon electrochemical biosensors for the detection of whole pathogenic bacteria. He is particularly interested in the optimisation and characterisation of biosensor surfaces at the nano-scale in order to maximise signal and minimise non-specific binding.

Paul A. Millner
After his BSc in Biochemistry, PhD in Plant Science at the University of Leeds (UK), then Postdoctoral Fellowships at Purdue University (Indiana, USA) and Imperial College (London, UK), Paul returned to Leeds in 1986 as a Lecturer. After 15 years as a plant biotechnologist/protein chemist, Paul moved into the area of nano- and bionanotechnology, with a particular interest in the development of biosensors for applications as diverse as medical diagnostics, environmental monitoring and detection of biological and chemical toxins. Paul is currently the Head of the School of Biomedical Sciences at the University of Leeds and also heads the Bionanotechnology Group. Current programmes in Paul's group include electrochemical biosensors for diagnosis of STIs, MRSA and other bacteria, biosensors for detecting bowel leakage after colorectal cancer resection; targeted and fluorescent nanoparticles for colorectal cancer; photosensitiser-loaded nanofibres for the solar sanitation of polluted water. Paul's work is united by a deep interest in bioengineering on the nanoscale by interfacing biological reagents with surfaces to result in electrical communication or enhanced activity.

References

(2013a). ABBOT LABORATORIES, Abbot Point Of Care.

(2013b). UNISCAN INSTRUMENTS, PG581 Potentiostat - Galvanostat.

Albrecht, C., Kaeppel, N., and Gauglitz, G. (2008). Two immunoassay formats for fully automated CRP detection in human serum. Anal Bioanal Chem *391*, 1845–1852.

Alizadeh, T., and Akbari, A. (2013). A capacitive biosensor for ultra-trace level urea determination based on nano-sized urea-imprinted polymer receptors coated on graphite electrode surface. Biosensors and Bioelectronics *43*, 321–327.

Andersson, K., Hämäläinen, M., and Malmqvist, M. (1999). Identification and optimization of regeneration conditions for affinity-based biosensor assays. A multivariate cocktail approach. Analytical Chemistry *71*, 2475–2481.

Arya, S.K., Chornokur, G., Venugopal, M., and Bhansali, S. (2010). Antibody functionalized interdigitated [small mu]-electrode (ID[small mu]E) based impedimetric cortisol biosensor. Analyst *135*, 1941–1946.

Arya, S.K., Dey, A., and Bhansali, S. (2011). Polyaniline protected gold nanoparticles based mediator and label free electrochemical cortisol biosensor. Biosensors and Bioelectronics *28*, 166–173.

Arya, S.K., Pui, T.S., Wong, C.C., Kumar, S., and Rahman, A.R.A. (2013). Effects of electrode size and modification protocol on label free electrochemical biosensor. Langmuir: the ACS Journal of Surfaces and Colloids.

Bain, C.D., Troughton, E.B., Tao, Y.T., Evall, J., Whitesides, G.M., and Nuzzo, R.G. (1989). Formation of monolayer films by the spontaneous assembly of organic thiols from solution onto gold. J Am Chem Soc *111*, 321–335.

Baldwin, E.A., Bai, J.H., Plotto, A., and Dea, S. (2011). Electronic noses and tongues: applications for the food and pharmaceutical industries. Sensors-Basel *11*, 4744–4766.

Barreiros dos Santos, M., Agusil, J.P., Prieto-Simón, B., Sporer, C., Teixeira, V., and Samitier, J. (2013). Highly sensitive detection of pathogen *Escherichia coli* O157:H7 by electrochemical impedance spectroscopy. Biosensors and Bioelectronics *45*, 174–180.

Barton, A.C., Collyer, S.D., Davis, F., Garifallou, G.Z., Tsekenis, G., Tully, E., O'Kennedy, R., Gibson, T., Millner, P.A., and Higson, S.P. (2009). Labeless AC impedimetric antibody-based sensors with pgml(-1) sensitivities for point-of-care biomedical applications. Biosens Bioelectron *24*, 1090–1095.

Berggren, C., Bjarnason, B., and Johansson, G. (2001). Capacitive biosensors. Electroanalysis *13*, 173–180.

Billah, M., Hays, H.C.W., and Millner, P.A. (2008). Development of a myoglobin impedimetric immunosensor based on mixed self-assembled monolayer onto gold. Microchim Acta *160*, 447–454.

Billah, M.M., Hodges, C.S., Hays, H.C., and Millner, P.A. (2010). Directed immobilization of reduced antibody fragments onto a novel SAM on gold for myoglobin impedance immunosensing. Bioelectrochemistry *80*, 49–54.

Bonanni, A., Loo, A.H., and Pumera, M. (2012). Graphene for impedimetric biosensing. Trac-Trend Anal Chem *37*, 12–21.

Bonroy, K., Frederix, F., Reekmans, G., Dewolf, E., De Palma, R., Borghs, G., Declerck, P., and Goddeeris, B. (2006). Comparison of random and oriented immobilisation of antibody fragments on mixed self-assembled monolayers. Journal of Immunological Methods *312*, 167–181.

Bourigua, S., Hnaien, M., Bessueille, F., Lagarde, F., Dzyadevych, S., Maaref, A., Bausells, J., Errachid, A., and Renault, N.J. (2010). Impedimetric immunosensor based on SWCNT-COOH modified gold microelectrodes for label-free detection of deep venous thrombosis biomarker. Biosensors and Bioelectronics *26*, 1278–1282.

Bright, F.V., Betts, T.A., and Litwiler, K.S. (1990). Regenerable fiber-optic-based immunosensor. Analytical Chemistry *62*, 1065–1069.

Bryan, T., Luo, X., Bueno, P.R., and Davis, J.J. (2013). An optimised electrochemical biosensor for the label-free detection of C-reactive protein in blood. Biosensors and Bioelectronics *39*, 94–98.

Castagnola, M., Picciotti, P.M., Messana, I., Fanali, C., Fiorita, A., Cabras, T., Calo, L., Pisano, E., Passali, G.C., Iavarone, F. et al. (2011). Potential applications of human saliva as diagnostic fluid. Acta Otorhinolaryngol Ital *31*, 347–357.

Cavalcanti, I.T., Guedes, M.I.F., Sotomayor, M.D.P.T., Yamanaka, H., and Dutra, R.F. (2012). A label-free immunosensor based on recordable compact disk chip for early diagnostic of the dengue virus infection. Biochem Eng J *67*, 225–230.

Caygill, R.L., Blair, G.E., and Millner, P.A. (2010). A review on viral biosensors to detect human pathogens. Anal Chim Acta *681*, 8–15.

Caygill, R.L., Hodges, C.S., Holmes, J.L., Higson, S.P., Blair, G.E., and Millner, P.A. (2012). Novel impedimetric immunosensor for the detection and quantitation of Adenovirus using reduced antibody fragments immobilized onto a conducting copolymer surface. Biosens Bioelectron *32*, 104–110.

Chang, B.Y., and Park, S.M. (2010). Electrochemical impedance spectroscopy. Annu Rev Anal Chem (Palo Alto Calif) *3*, 207–229.

Chauhan, N., Narang, J., Sunny, and Pundir, C.S. (2013). Immobilization of lysine oxidase on a gold–platinum nanoparticles modified Au electrode for detection of lysine. Enzyme Microb Tech *52*, 265–271.

Chen, Q., Tang, W., Wang, D., Wu, X., Li, N., and Liu, F. (2010). Amplified QCM-D biosensor for protein based on aptamer-functionalized gold nanoparticles. Biosensors and Bioelectronics *26*, 575–579.

Choi, S., Goryll, M., Sin, L., Wong, P., and Chae, J. (2011). Microfluidic-based biosensors toward point-of-care detection of nucleic acids and proteins. Microfluidics and Nanofluidics *10*, 231–247.

Chowdhury, A.D., De, A., Chaudhuri, C.R., Bandyopadhyay, K., and Sen, P. (2012). Label free polyaniline based impedimetric biosensor for detection of E. coli O157:H7 Bacteria. Sensor Actuat B-Chem *171*, 916–923.

Chuang, Y.-H., Chang, Y.-T., Liu, K.-L., Chang, H.-Y., and Yew, T.-R. (2011). Electrical impedimetric biosensors for liver function detection. Biosensors and Bioelectronics *28*, 368–372.

Chullasat, K., Kanatharana, P., Limbut, W., Numnuam, A., and Thavarungkul, P. (2011). Ultra trace analysis of small molecule by label-free impedimetric immunosensor using multilayer modified electrode. Biosensors and Bioelectronics *26*, 4571–4578.

Clark, L.C., Jr., and Lyons, C. (1962). Electrode systems for continuous monitoring in cardiovascular surgery. Ann N Y Acad Sci 102, 29–45.

Cole, K.S. (1928). Electric impedance of suspensions of Arbacia eggs. J Gen Physiol 12, 37–54.

Cole, K.S. (1932). Electric phase angle of cell mambranes. J Gen Physiol 15, 641–649.

Connolly, P. (2004). The potential for biosensors in cardiac surgery. Perfusion-UK 19, 247–249.

Conroy, D.J., Millner, P.A., Stewart, D.I., and Pollmann, K. (2010). Biosensing for the environment and defence: aqueous uranyl detection using bacterial surface layer proteins. Sensors (Basel) 10, 4739–4755.

Conroy, P.J., Hearty, S., Leonard, P., and O'Kennedy, R.J. (2009). Antibody production, design and use for biosensor-based applications. Seminars in Cel&; Developmental Biology 20, 10–26.

Cooper, M.A. (2002). Optical biosensors in drug discovery. Nature Reviews Drug Discovery 1, 515–528.

Corry, B., Uilk, J., and Crawley, C. (2003). Probing direct binding affinity in electrochemical antibody-based sensors. Analytica Chimica Acta 496, 103–116.

Cosnier, S. (2005). Affinity biosensors based on electropolymerized films. Electroanalysis 17, 1701–1715.

Daniels, J.S., and Pourmand, N. (2007). Label-free impedance biosensors: opportunities and challenges. Electroanalysis 19, 1239–1257.

Das, M., Dhand, C., Sumana, G., Srivastava, A.K., Vijayan, N., Nagarajan, R., and Malhotra, B.D. (2011). Zirconia grafted carbon nanotubes based biosensor for M. Tuberculosis detection. Appl Phys Lett 99.

De Volder, M.F.L., Tawfick, S.H., Baughman, R.H., and Hart, A.J. (2013). Carbon nanotubes: present and future commercial applications. Science 339, 535–539.

Dillon, P.P., Daly, S.J., Manning, B.M., and O'Kennedy, R. (2003). Immunoassay for the determination of morphine-3-glucuronide using a surface plasmon resonance-based biosensor. Biosensors and Bioelectronics 18, 217–227.

Dillon, P.P., Killard, A.J., Daly, S.J., Leonard, P., and O'Kennedy, R. (2005). Novel assay format permitting the prolonged use of regeneration-based sensor chip technology. Journal of Immunological Methods 296, 77–82.

Dinçkaya, E., Kınık, Ö., Sezgintürk, M.K., Altuğ, Ç., and Akkoca, A. (2011). Development of an impedimetric aflatoxin M1 biosensor based on a DNA probe and gold nanoparticles. Biosensors and Bioelectronics 26, 3806–3811.

Drake, A.W., and Klakamp, S.L. (2011). A strategic and systematic approach for the determination of biosensor regeneration conditions. Journal of Immunological Methods 371, 165–169.

Dubuisson, M. (1937). Impedance changes in muscle during contraction, and their possible relation to chemical processes. J Physiol 89, 132–152.

Eggins, B.R. (2002). Chemical sensors and biosensors. Chinchester: John Wiley & Sons.

Elshafey, R., Tlili, C., Abulrob, A., Tavares, A.C., and Zourob, M. (2013). Label-free impedimetric immunosensor for ultrasensitive detection of cancer marker Murine double minute 2 in brain tissue. Biosensors and Bioelectronics 39, 220–225.

Erickson, K.A., and Wilding, P. (1993). Evaluation of a novel point-of-care system, the i-stat portable clinical analyzer. Clin Chem 39, 283–287.

Escamilla-Gómez, V., Campuzano, S., Pedrero, M., and Pingarrón, J.M. (2009). Gold screen-printed-based impedimetric immunobiosensors for direct and sensitive *Escherichia coli* quantisation. Biosensors and Bioelectronics 24, 3365–3371.

Esseghaier, C., Ng, A., and Zourob, M. (2013). A novel and rapid assay for HIV-1 protease detection using magnetic bead mediation. Biosens Bioelectron 41, 335–341.

Fan, X., White, I.M., Shopova, S.I., Zhu, H., Suter, J.D., and Sun, Y. (2008). Sensitive optical biosensors for unlabeled targets: a review. Analytica Chimica Acta 620, 8–26.

Fenter, P., Eberhardt, A., and Eisenberger, P. (1994). Self-Assembly of n-Alkyl Thiols as Disulfides on Au(111). Science 266, 1216–1218.

Flynn, N.T., Tran, T.N.T., Cima, M.J., and Langer, R. (2003). Long-term stability of self-assembled monolayers in biological media. Langmuir 19, 10909–10915.

Fricke, H. (1925). The electric capacity of suspensions with special reference to blood. J Gen Physiol 9, 137–152.

Gamella, M., Campuzano, S., Parrado, C., Reviejo, A.J., and Pingarron, J.M. (2009). Microorganisms recognition and quantification by lectin adsorptive affinity impedance. Talanta 78, 1303–1309.

Geddes, L.A., and Roeder, R. (2003). Criteria for the selection of materials for implanted electrodes. Ann Biomed Eng 31, 879–890.

Geng, P., Zhang, X., Meng, W., Wang, Q., Zhang, W., Jin, L., Feng, Z., and Wu, Z. (2008). Self-assembled monolayers-based immunosensor for detection of *Escherichia coli* using electrochemical impedance spectroscopy. Electrochim Acta 53, 4663–4668.

Gerard, M., Chaubey, A., and Malhotra, B.D. (2002). Application of conducting polymers to biosensors. Biosensors & Bioelectronics 17, 345–359.

Gilbreth, R.N., and Koide, S. (2012). Structural insights for engineering binding proteins based on non-antibody scaffolds. Current Opinion in Structural Biology 22, 413–420.

Grieshaber, D., MacKenzie, R., Voros, J., and Reimhult, E. (2008). Electrochemical biosensors - sensor principles and architectures. Sensors-Basel 8, 1400–1458.

Guan, J.G., Miao, Y.Q., and Zhang, Q.J. (2004). Impedimetric biosensors. Journal of Bioscience and Bioengineering 97, 219–226.

Hakkinen, H. (2012). The gold-sulfur interface at the nanoscale. Nat Chem 4, 443–455.

Hall, H.P. (2004). How electronics changed impedance measurements. Paper presented at: Instrumentation and Measurement Technology Conference, 2004 IMTC 04 Proceedings of the 21st IEEE.

Hassen, W.M., Duplan, V., Frost, E., and Dubowski, J.J. (2011). Quantitation of influenza A virus in the presence of extraneous protein using electrochemical impedance spectroscopy. Electrochim Acta 56, 8325–8328.

Hays, H.C.W., Millner, P.A., and Prodromidis, M.I. (2006). Development of capacitance based immunosensors on mixed self-assembled monolayers. Sensors and Actuators B: Chemical 114, 1064–1070.

He, F.J., Zhao, J.W., Zhang, L.D., and Su, X.N. (2003). A rapid method for determining *Mycobacterium tuberculosis* based on a bulk acoustic wave impedance biosensor. Talanta 59, 935–941.

Heaviside, O. (1894a). Electrical Papers, Volume 1. New York: MacMillan.

Heaviside, O. (1894b). Electrical Papers, Volume 2. New York: MacMillan.

Hermanson, G.T. (2008). Bioconjugate Techniques, 2nd Edition. Bioconjugate Techniques, 2nd Edition, 1–1202.

Hiatt, L.A., and Cliffel, D.E. (2012). Real-time recognition of *Mycobacterium tuberculosis* and lipoarabinomannan using the quartz crystal microbalance. Sensor Actuat B-Chem 174, 245–252.

Hock, B., Seifert, M., and Kramer, K. (2002). Engineering receptors and antibodies for biosensors. Biosens Bioelectron 17, 239–249.

Holliger, P., and Hudson, P.J. (2005). Engineered antibody fragments and the rise of single domains. Nature Biotechnology 23, 1126–1136.

Homola, J., Yee, S.S., and Gauglitz, G. (1999). Surface plasmon resonance sensors: review. Sens Actuator B-Chem 54, 3–15.

Hong, S.-R., Choi, S.-J., Jeong, H.D., and Hong, S. (2009). Development of QCM biosensor to detect a marine derived pathogenic bacteria Edwardsiella tarda using a novel immobilisation method. Biosensors and Bioelectronics 24, 1635–1640.

Hu, Y., Li, F., Han, D., Wu, T., Zhang, Q., Niu, L., and Bao, Y. (2012). Simple and label-free electrochemical assay for signal-on DNA hybridization directly at undecorated graphene oxide. Analytica Chimica Acta 753, 82–89.

Hu, Y., Zuo, P., and Ye, B.-C. (2013). Label-free electrochemical impedance spectroscopy biosensor for direct detection of cancer cells based on the interaction between carbohydrate and lectin. Biosensors and Bioelectronics 43, 79–83.

Huang, J., Yang, G., Meng, W., Wu, L., Zhu, A., and Jiao, X.a. (2010). An electrochemical impedimetric immunosensor for label-free detection of Campylobacter jejuni in diarrhea patients' stool based on O-carboxymethylchitosan surface modified Fe3O4 nanoparticles. Biosensors and Bioelectronics 25, 1204–1211.

Huang, Y., and Suni, I.I. (2008). Degenerate Si as an electrode material for electrochemical biosensors. J Electrochem Soc 155, J350–J354.

Hughes, M.D. (2009). The business of self-monitoring of blood glucose: a market profile. Journal of diabetes science and technology 3, 1219–1223.

Ivanov, Y.D., Pleshakova, T.O., Kozlov, A.F., Malsagova, K.A., Krohin, N.V., Shumyantseva, V.V., Shumov, I.D., Popov, V.P., Naumova, O.V., Fomin, B.I., et al. (2012). SOI nanowire for the high-sensitive detection of HBsAg and [small alpha]-fetoprotein. Lab on a chip 12, 5104–5111.

Janshoff, A., Galla, H.J., and Steinem, C. (2000). Piezoelectric mass-sensing devices as biosensors - An alternative to optical biosensors? Angewandte Chemie-International Edition 39, 4004–4032.

Joung, C.-K., Kim, H.-N., Lim, M.-C., Jeon, T.-J., Kim, H.-Y., and Kim, Y.-R. (2013). A nanoporous membrane-based impedimetric immunosensor for label-free detection of pathogenic bacteria in whole milk. Biosensors and Bioelectronics 44, 210–215.

Kadara, R.O., Jenkinson, N., and Banks, C.E. (2009). Characterisation of commercially available electrochemical sensing platforms. Sensors and Actuators B: Chemical 138, 556–562.

Kandimalla, V.B., Neeta, N.S., Karanth, N.G., Thakur, M.S., Roshini, K.R., Rani, B.E.A., Pasha, A., and Karanth, N.G.K. (2004). Regeneration of ethyl parathion antibodies for repeated use in immunosensor: a study on dissociation of antigens from antibodies. Biosensors and Bioelectronics 20, 903–906.

Karyakin, A.A., Gitelmacher, O.V., and Karyakina, E.E. (1995). Prussian blue based first-generation biosensor - a sensitive amperometric electrode for glucose. Anal Chem 67, 2419–2423.

Kashefi-Kheyrabadi, L., and Mehrgardi, M.A. (2012). Design and construction of a label free aptasensor for electrochemical detection of sodium diclofenac. Biosensors and Bioelectronics 33, 184–189.

Katz, E., and Willner, I. (2003). Probing biomolecular interactions at conductive and semiconductive surfaces by impedance spectroscopy: routes to impedimetric immunosensors, DNA-Sensors, and enzyme biosensors. Electroanalysis 15, 913–947.

Kiilerich-Pedersen, K., and Rozlosnik, N. (2012). Cell-based biosensors: electrical sensing in microfluidic devices. Diagnostics 2, 83–96.

Kim, B.K., Li, J., Im, J.E., Ahn, K.S., Park, T.S., Cho, S.I., Kim, Y.R., and Lee, W.Y. (2012). Impedometric estrogen biosensor based on estrogen receptor alpha-immobilized gold electrode. J Electroanal Chem 671, 106–111.

Kim, H., Namgung, R., Singha, K., Oh, I.-K., and Kim, W.J. (2011). Graphene oxide–polyethylenimine nanoconstruct as a gene delivery vector and bioimaging tool. Bioconjugate chemistry 22, 2558–2567.

Komarova, E., Reber, K., Aldissi, M., and Bogomolova, A. (2010). New multispecific array as a tool for electrochemical impedance spectroscopy-based biosensing. Biosensors & Bioelectronics 25, 1389–1394.

Korotcenkov, G. (2010). Chemical Sensors: Volume 1 General Approaches (Momentum Press).

Kotanen, C.N., Moussy, F.G., Carrara, S., and Guiseppi-Elie, A. (2012). Implantable enzyme amperometric biosensors. Biosensors and Bioelectronics 35, 14–26.

Kurkina, T., Vlandas, A., Ahmad, A., Kern, K., and Balasubramanian, K. (2011). Label-free detection of few copies of DNA with carbon nanotube impedance biosensors. Angewandte Chemie International Edition 50, 3710–3714.

Lavrik, N.V., Sepaniak, M.J., and Datskos, P.G. (2004). Cantilever transducers as a platform for chemical and biological sensors. Review of Scientific Instruments 75, 2229–2253.

Lazerges, M., Perrot, H., Zeghib, N., Antoine, E., and Compere, C. (2006). In situ QCM DNA-biosensor probe modification. Sensors and Actuators B: Chemical 120, 329–337.

Lee, J.A., Hwang, S., Kwak, J., Il Park, S., Lee, S.S., and Lee, K.C. (2008). An electrochemical impedance biosensor with aptamer-modified pyrolyzed carbon electrode for label-free protein detection. Sensor Actuat B-Chem 129, 372–379.

Lee, J.A., Hwang, S., Lee, K.C., Kwak, J., Park, S.I., and Lee, S.S. (2007). Pyrolyzed carbon biosenosor for aptamer-protein interactions using electrochemical impedance spectroscopy. Proc Ieee Micr Elect, 87–90.

Lemaitre, L., Moors, M., and van Peterghem, A.P. (1985). AC impedance measurements on high copper dental amalgams. Biomaterials 6, 425–426.

Liu, Q., Yu, J., Xiao, L., Tang, J.C.O., Zhang, Y., Wang, P., and Yang, M. (2009). Impedance studies of bio-behavior and chemosensitivity of cancer cells by micro-electrode arrays. Biosensors and Bioelectronics 24, 1305–1310.

Loo, A.H., Bonanni, A., and Pumera, M. (2012). Impedimetric thrombin aptasensor based on chemically modified graphenes. Nanoscale 4, 143–147.

Lu, B., Smyth, M.R., and O'Kennedy, R. (1996). Oriented immobilization of antibodies and its applications in immunoassays and immunosensors. The Analyst 121, 29R–32R.

Lu, L., Chee, G., Yamada, K., and Jun, S. (2013). Electrochemical impedance spectroscopic technique with a functionalized microwire sensor for rapid detection of foodbornepathogens. Biosensors and Bioelectronics 42, 492–495.

Luong, J.H.T., Male, K.B., and Glennon, J.D. (2008). Biosensor technology: Technology push versus market pull. Biotechnology Advances 26, 492–500.

Luppa, P.B., Sokoll, L.J., and Chan, D.W. (2001). Immunosensors - principles and applications to clinical chemistry. Clin Chim Acta 314, 1–26.

Macdonald, D.D. (2006). Reflections on the history of electrochemical impedance spectroscopy. Electrochim Acta 51, 1376–1388.

Malhotra, B.D., and Chaubey, A. (2003). Biosensors for clinical diagnostics industry. Sensors and Actuators B: Chemical 91, 117–127.

Manso, J., Mena, M.L., Yáñez-Sedeño, P., and Pingarrón, J.M. (2008). Alcohol dehydrogenase amperometric biosensor based on a colloidal gold–carbon nanotubes composite electrode. Electrochimica Acta 53, 4007–4012.

Mantzila, A.G., Maipa, V., and Prodromidis, M.I. (2008). Development of a faradic impedimetric immunosensor for the detection of Salmonella typhimurium in milk. Anal Chem 80, 1169–1175.

Mattos, A.B., Freitas, T.A., Silva, V.L., and Dutra, R.F. (2012). A dual quartz crystal microbalance for human cardiac troponin T in real time detection. Sensors and Actuators B: Chemical 161, 439–446.

McNerney, R., and Daley, P. (2011). Towards a point-of-care test for active tuberculosis: obstacles and opportunities. Nat Rev Microbiol 9, 204–213.

Mejri, M.B., Baccar, H., Baldrich, E., Del Campo, F.J., Helali, S., Ktari, T., Simonian, A., Aouni, M., and Abdelghani, A. (2010). Impedance biosensing using phages for bacteria detection: generation of dual signals as the clue for in-chip assay confirmation. Biosensors and Bioelectronics 26, 1261–1267.

Metters, J.P., Kadara, R.O., and Banks, C.E. (2011). New directions in screen printed electroanalytical sensors: an overview of recent developments. The Analyst 136, 1067–1076.

Meyer, S.C., and Ghosh, I. (2010). Phage Display Technology in Biosensor Development.

Michalzik, M., Wendler, J., Rabe, J., Büttgenbach, S., and Bilitewski, U. (2005). Development and application of a miniaturised quartz crystal microbalance (QCM) as immunosensor for bone morphogenetic protein-2. Sensors and Actuators B: Chemical 105, 508–515.

Millan, K.M., and Mikkelsen, S.R. (1993). Sequence-selective biosensor for DNA-based on electroactive hybridization indicators. Anal Chem 65, 2317–2323.

Millner, P.A., Caygill, R.L., Conroy, D.J., and Shahidan, M.A. (2012). Impedance interrogated affinity biosensors for medical applications: novel targets and mechanistic studies, Vol. 45 (Cambridge, UK: Woodhead Publishing Ltd.).

Millner, P.A., Hays, H.C., Vakurov, A., Pchelintsev, N.A., Billah, M.M., and Rodgers, M.A. (2009). Nanostructured transducer surfaces for electrochemical biosensor construction--interfacing the sensing component with the electrode. Seminars in cell & developmental biology 20, 34–40.

Mir, M., Homs, A., and Samitier, J. (2009). Integrated electrochemical DNA biosensors for lab-on-a-chip devices. Electrophoresis 30, 3386–3397.

Miscoria, S.A., Barrera, G.D., and Rivas, G.A. (2006). Glucose biosensors based on the immobilization of glucose oxidase and polytyramine on rodhinized glassy carbon and screen printed electrodes. Sens Actuator B-Chem 115, 205–211.

Muramatsu, H., Dicks, J.M., Tamiya, E., and Karube, I. (1987). Piezoelectric crystal biosensor modified with protein-A for determination of immunoglobulins. Anal Chem 59, 2760–2763.

Newman, A.L., Hunter, K.W., and Stanbro, W.D. (1986). The capacitive affinity sensor: a new biosensor. Chemical Sensors: 2nd International Meeting, Proc, 596–598.

Newman, J.D., and Turner, A.P.F. (2005). Home blood glucose biosensors: a commercial perspective. Biosensors & Bioelectronics 20, 2435–2453.

Ng, S.Y., Reboud, J., Wang, K.Y.P., Tang, K.C., Zhang, L., Wong, P., Moe, K.T., Shim, W., and Chen, Y. (2010). Label-free impedance detection of low levels of circulating endothelial progenitor cells for point-of-care diagnosis. Biosensors and Bioelectronics 25, 1095–1101.

Novoselov, K.S., Geim, A.K., Morozov, S.V., Jiang, D., Zhang, Y., Dubonos, S.V., Grigorieva, I.V., and Firsov, A.A. (2004). Electric field effect in atomically thin carbon films. Science 306, 666–669.

Ohno, R., Ohnuki, H., Wang, H., Yokoyama, T., Endo, H., Tsuya, D., and Izumi, M. (2013). Electrochemical impedance spectroscopy biosensor with interdigitated electrode for detection of human immunoglobulin A. Biosensors and Bioelectronics 40, 422–426.

Oliveira, M.D.L., Nogueira, M.L., Correia, M.T.S., Coelho, L.C.B.B., and Andrade, C.A.S. (2011). Detection of dengue virus serotypes on the surface of gold electrode based on Cratylia mollis lectin affinity. Sensor Actuat B-Chem 155, 789–795.

Pang, P.F., Cai, Q.Y., Yao, S.Z., and Grimes, C.A. (2008). The detection of *Mycobacterium tuberculosis* in sputum sample based on a wireless magnetoelastic-sensing device. Talanta 76, 360–364.

Parsajoo, C., Kauffmann, J.M., and Elkaoutit, M. (2012). Biosensors for drug testing and discovery. In Biosensors for medical applications, S.P. Higson, ed. (Cambridge, UK.: Woodhead Publishing Ltd.), pp. 233–262.

Patolsky, F., Zheng, G., and Lieber, C.M. (2006). Nanowire sensors for medicine and the life sciences. Nanomedicine (Lond) 1, 51–65.

Pchelintsev, N.A., and Millner, P.A. (2008). A novel procedure for rapid surface functionalisation and mediator loading of screen-printed carbon electrodes. Anal Chim Acta 612, 190–197.

Pchelintsev, N.A., Vakurov, A., and Millner, P.A. (2009). Simultaneous deposition of Prussian Blue and creation of an electrostatic surface for rapid biosensor construction. Sensors and Actuators B: Chemical 138, 461–466.

Peh, A.E.K., and Li, S.F.Y. (2013). Dengue virus detection using impedance measured across nanoporous aluminamembrane. Biosensors and Bioelectronics 42, 391–396.

Pike, D.J., Kapur, N., Millner, P.A., and Stewart, D.I. (2013). Flow cell design for effective biosensing. Sensors (Switzerland) 13, 58–70.

Piro, B., Reisberg, S., Noel, V., and Pham, M.C. (2007). Investigations of the steric effect on electrochemical transduction in a quinone-based DNA sensor. Biosensors and Bioelectronics 22, 3126–3131.

Pohanka, M., and Skladai, P. (2008). Electrochemical biosensors - principles and applications. J Appl Biomed 6, 57–64.

Pournaras, A.V., Koraki, T., and Prodromidis, M.I. (2008). Development of an impedimetric immunosensor based on electropolymerized polytyramine films for the direct detection of Salmonella typhimurium in pure cultures of type strains and inoculated real samples. Analytica Chimica Acta 624, 301–307.

Prodromidis, M.I. (2010). Impedimetric immunosensors—A review. Electrochimica Acta 55, 4227–4233.

Qi, P., Wan, Y., and Zhang, D. (2013). Impedimetric biosensor based on cell-mediated bioimprinted films for bacterial detection. Biosens Bioelectron 39, 282–288.

Queirós, R.B., de-los-Santos-Álvarez, N., Noronha, J.P., and Sales, M.G.F. (2013). A label-free DNA aptamer-based impedance biosensor for the detection of E. coli outer membrane proteins. Sensors and Actuators B: Chemical 181, 766–772.

Radi, A.-E., Acero Sánchez, J.L., Baldrich, E., and O'Sullivan, C.K. (2005). Reagentless, reusable, ultrasensitive electrochemical molecular beacon aptasensor. Journal of the American Chemical Society 128, 117–124.

Raffa, D.L., Leung, K.T., and Battaglini, F. (2006). Electrochemical copolymerization of aniline and ortho-aminobenzylamine. Studies on its conductivity and chemical derivatization. Journal of Electroanalytical Chemistry 587, 60–66.

Ramanavicius, A., Habermüller, K., Csöregi, E., Laurinavicius, V., and Schuhmann, W. (1999). Polypyrrole-entrapped quinohemoprotein alcohol dehydrogenase. evidence for direct electron transfer via conducting-polymer chains. Analytical Chemistry 71, 3581–3586.

Ramulu, T.S., Venu, R., Sinha, B., Lim, B., Jeon, S.J., Yoon, S.S., and Kim, C.G. (2013). Nanowires array modified electrode for enhanced electrochemical detection of nucleic acid. Biosensors and Bioelectronics 40, 258–264.

Randles, J.E.B. (1947). Kinetics of rapid electrode reactions. Discussions of the Faraday Society 1, 11–19.

Ravalli, A., dos Santos, G.P., Ferroni, M., Faglia, G., Yamanaka, H., and Marrazza, G. (2013). New label free CA125 detection based on gold nanostructured screen-printed electrode. Sensors and Actuators B: Chemical 179, 194–200.

Reddy, A.S.G., Narakathu, B.B., Atashbar, M.Z., Rebros, M., Rebrosova, E., and Joyce, M.K. (2011). Gravure printed electrochemical biosensor. Procedia Engineering 25, 956–959.

Renedo, O.D., Alonso-Lomillo, M.A., and Martínez, M.J.A. (2007). Recent developments in the field of screen-printed electrodes and their related applications. Talanta 73, 202–219.

Reyes, D.R., Iossifidis, D., Auroux, P.-A., and Manz, A. (2002). Micro total analysis systems. 1. Introduction, theory, and technology. Analytical Chemistry 74, 2623–2636.

Ricci, F., and Palleschi, G. (2005). Sensor and biosensor preparation, optimisation and applications of Prussian Blue modified electrodes. Biosensors & Bioelectronics 21, 389–407.

Rivet, C., Lee, H., Hirsch, A., Hamilton, S., and Lu, H. (2011). Microfluidics for medical diagnostics and biosensors. Chemical Engineering Science 66, 1490–1507.

Rodgers, M.A., Findlay, J.B.C., and Millner, P.A. (2010). Lipocalin based biosensors for low mass hydrophobic analytes; development of a novel SAM for polyhistidine tagged proteins. Sensor Actuat B-Chem 150, 12–18.

Roine, A., Tolvanen, M., Sipiläinen, M., Kumpulainen, P., Helenius, M.A., Lehtimäki, T., Vepsäläinen, J., Keinänen, T.A., Häkkinen, M.R., Koskimäki, J. et al. (2012). Detection of smell print differences between nonmalignant and malignant prostate cells with an electronic nose. Future Oncology 8, 1157–1165.

Ronkainen, N.J., Halsall, H.B., and Heineman, W.R. (2010). Electrochemical biosensors. Chemical Society Reviews 39, 1747–1763.

Sahoo, P., Suresh, S., Dhara, S., Saini, G., Rangarajan, S., and Tyagi, A.K. (2013). Direct label free ultrasensitive impedimetric DNA biosensor using dendrimer functionalized GaN nanowires. Biosensors and Bioelectronics 44, 164–170.

Scognamiglio, V. (2013). Nanotechnology in glucose monitoring: advances and challenges in the last 10 years. Biosensors and Bioelectronics 47, 12–25.

Scouten, W.H., Luong, J.H.T., and Stephen Brown, R. (1995). Enzyme or protein immobilization techniques for applications in biosensor design. Trends in Biotechnology 13, 178–185.

Sezgintürk, M.K. (2011). A new impedimetric biosensor utilizing VEGF receptor-1 (Flt-1): early diagnosis of vascular endothelial growth factor in breast cancer. Biosensors and Bioelectronics 26, 4032–4039.

Shabani, A., Zourob, M., Allain, B., Marquette, C.A., Lawrence, M.F., and Mandeville, R. (2008). Bacteriophage-modified microarrays for the direct impedimetric detection of bacteria. Anal Chem 80, 9475–9482.

Silva, G.J.L., Andrade, C.A.S., Oliveira, I.S., de Melo, C.P., and Oliveira, M.D.L. (2013). Impedimetric sensor for toxigenic Penicillium sclerotigenum detection in yam based on magnetite-poly(allylamine hydrochloride) composite. Journal of Colloid and Interface Science 396, 258–263.

Simmen, H.P., Battaglia, H., Giovanoli, P., and Blaser, J. (1994). Analysis of PH, PO(2) and PCO(2) in drainage fluid allows for rapid detection of infectious complications during the follow-up period after abdominal-surgery. Infection 22, 386–389.

Soledad Belluzo, M., Elida Ribone, M., Camussone, C., Sergio Marcipar, I., and Marina Lagier, C. (2011). Favorably orienting recombinant proteins to develop amperometric biosensors to diagnose Chagas' disease. Analytical Biochemistry 408, 86–94.

Song, S., Wang, L., Li, J., Zhao, J., and Fan, C. (2008). Aptamer-based biosensors. Trac-Trends in Analytical Chemistry 27, 108–117.

Song, S., Xu, H., and Fan, C. (2006). Potential diagnostic applications of biosensors: current and future directions. International Journal of Nanomedicine 1, 433–440.

Soper, S.A., Brown, K., Ellington, A., Frazier, B., Garcia-Manero, G., Gau, V., Gutman, S.I., Hayes, D.F., Korte, B., Landers, J.L. et al. (2006). Point-of-care biosensor systems for cancer diagnostics/prognostics. Biosensors and Bioelectronics 21, 1932–1942.

Steegborn, C., and Skládal, P. (1997). Construction and characterization of the direct piezoelectric immunosensor for atrazine operating in solution. Biosensors and Bioelectronics 12, 19–27.

Suni, I.I. (2008). Impedance methods for electrochemical sensors using nanomaterials. Trac-Trend Anal Chem 27, 604–611.

Svancara, I., Vytras, K., Kalcher, K., Walcarius, A., and Wang, J. (2009). Carbon paste electrodes in facts, numbers, and notes: a review on the occasion of the 50-years Jubilee of carbon paste in electrochemistry and electroanalysis. Electroanalysis 21, 7–28.

Tägil, M., Geijer, M., Abramo, A., and Kopylov, P. (2013). Ten years' experience with a pyrocarbon prosthesis replacing the proximal interphalangeal joint. A prospective clinical and radiographic follow-up. Journal of Hand Surgery (European Volume).

Tan, F., Leung, P.H.M., Liu, Z.B., Zhang, Y., Xiao, L.D., Ye, W.W., Zhang, X., Yi, L., and Yang, M. (2011). A PDMS microfluidic impedance immunosensor for E. coli O157:H7 and Staphylococcus aureus detection via antibody-immobilized nanoporous membrane. Sensor Actuat B-Chem 159, 328–335.

Taylor, R.F., Marenchic, I.G., and Cook, E.J. (1988). An Acetylcholine receptor-based biosensor for the detection of cholinergic agents. Analytica Chimica Acta 213, 131–138.

Tedeschi, L., Citti, L., and Domenici, C. (2005). An integrated approach for the design and synthesis of oligonucleotide probes and their interfacing to a QCM-based RNA biosensor. Biosensors and Bioelectronics 20, 2376–2385.

Tiefenauer, L., and Ros, R. (2002). Biointerface analysis on a molecular level: New tools for biosensor research. Colloids and Surfaces B: Biointerfaces 23, 95–114.

Trojanowicz, M., and Krawczyński vel Krawczyk, T. (1995). Electrochemical biosensors based on enzymes immobilized in electropolymerized films. Mikrochim Acta 121, 167–181.

Tsekenis, G., Garifallou, G.Z., Davis, F., Millner, P.A., Pinacho, D.G., Sanchez-Baeza, F., Marco, M.P., Gibson, T.D., and Higson, S.P. (2008). Detection of fluoroquinolone antibiotics in milk via a labeless immunoassay based upon an alternating current impedance protocol. Anal Chem 80, 9233–9239.

Turner, A.P.F. (1997). Biosensors: past, present and future.

Vakurov, A., Simpson, C.E., Daly, C.L., Gibson, T.D., and Millner, P.A. (2005). Acetylecholinesterase-based biosensor electrodes for organophosphate pesticide detection. II. Immobilization and stabilization of acetylecholinesterase. Biosens Bioelectron 20, 2324–2329.

van Noort, D., and Mandenius, C.-F. (2000). Porous gold surfaces for biosensor applications. Biosensors and Bioelectronics 15, 203–209.

Vidal, J.-C., Espuelas, J., and Castillo, J.-R. (2004). Amperometric cholesterol biosensor based on in situ reconstituted cholesterol oxidase on an immobilized monolayer of flavin adenine dinucleotide cofactor. Analytical Biochemistry 333, 88–98.

Vo-Dinh, T., and Cullum, B. (2000). Biosensors and biochips: advances in biological and medical diagnostics. Fresenius' journal of analytical chemistry 366, 540–551.

Wang, J. (2000). From DNA biosensors to gene chips. Nucleic Acids Research 28, 3011–3016.

Wang, J. (2001). Glucose biosensors: 40 years of advances and challenges. Electroanalysis 13, 983–988.

Wang, J. (2008). Electrochemical glucose biosensors. Chem Rev 108, 814–825.

Wang, M.J., Wang, L.Y., Wang, G., Ji, X.H., Bai, Y.B., Li, T.J., Gong, S.Y., and Li, J.H. (2004). Application of impedance spectroscopy for monitoring colloid Au-enhanced antibody immobilization and antibody-antigen reactions. Biosensors & Bioelectronics 19, 575–582.

Wang, R., Ruan, C., Kanayeva, D., Lassiter, K., and Li, Y. (2008). TiO2 nanowire bundle microelectrode based impedance immunosensor for rapid and sensitive detection of listeria monocytogenes. Nano Letters 8, 2625–2631.

Wang, S., Inci, F., De Libero, G., Singhal, A., and Demirci, U. (2013a). Point-of-care assays for tuberculosis: Role of nanotechnology/microfluidics. Biotechnology Advances 31, 438–449.

Wang, Y., Papadimitrakopoulos, F., and Burgess, D.J. (2013b). Polymeric "smart" coatings to prevent foreign body response to implantable biosensors. Journal of Controlled Release.

Wang, Y.X., Ye, Z.Z., and Ying, Y.B. (2012). New trends in impedimetric biosensors for the detection of foodborne pathogenic bacteria. Sensors-Basel 12, 3449–3471.

Whitesides, G.M. (2006). The origins and the future of microfluidics. Nature 442, 368–373.

Wicklein, B., del Burgo, M.A.M., Yuste, M., Carregal-Romero, E., Llobera, A., Darder, M., Aranda, P., Ortin, J., del Real, G., Fernandez-Sanchez, C. et al. (2013). Biomimetic architectures for the impedimetric discrimination of influenza virus phenotypes. Adv Funct Mater 23, 254–262.

Wijesuriya, D., Breslin, K., Anderson, G., Shriver-Lake, L., and Ligler, F.S. (1994). Regeneration of immobilized antibodies on fiber optic probes. Biosensors and Bioelectronics 9, 585–592.

Wilson, A. (2013). Diverse Applications of Electronic-Nose Technologies in Agriculture and Forestry. Sensors-Basel 13, 2295–2348.

Wilson, A.D., and Baietto, M. (2011). Advances in Electronic-Nose Technologies Developed for Biomedical Applications. Sensors-Basel 11, 1105–1176.

Wilson, G.S., and Gifford, R. (2005). Biosensors for real-time in vivo measurements. Biosensors and Bioelectronics 20, 2388–2403.

Wilson, R., and Turner, A.P.F. (1992). Glucose oxidase - an ideal enzyme. Biosens Bioelectron 7, 165–185.

Wink, T., J. van Zuilen, S., Bult, A., and P. van Bennekom, W. (1997). Self-assembled monolayers for biosensors. The Analyst 122, 43R–50R.

Xiao, Y., Lubin, A.A., Heeger, A.J., and Plaxco, K.W. (2005). Label-free electronic detection of thrombin in blood serum by using an aptamer-based sensor. Angewandte Chemie-International Edition 44, 5456–5459.

Xu, M., Luo, X., and Davis, J.J. (2013). The label free picomolar detection of insulin in blood serum. Biosensors and Bioelectronics 39, 21–25.

Yager, P., Domingo, G.J., and Gerdes, J. (2008). Point-of-care diagnostics for global health. In Annual Review of Biomedical Engineering (Palo Alto: Annual Reviews), pp. 107–144.

Yang, H., Li, Z., Wei, X., Huang, R., Qi, H., Gao, Q., Li, C., and Zhang, C. (2013a). Detection and discrimination of alpha-fetoprotein with a label-free electrochemical impedance spectroscopy biosensor array based on lectin functionalized carbon nanotubes. Talanta 111, 62–68.

Yang, L., and Bashir, R. (2008). Electrical/electrochemical impedance for rapid detection of foodborne pathogenic bacteria. Biotechnol Adv 26, 135–150.

Yang, T., Guo, X., Ma, Y., Li, Q., Zhong, L., and Jiao, K. (2013b). Electrochemical impedimetric DNA sensing based on multi-walled carbon nanotubes–SnO2–chitosan nanocomposite. Colloids and Surfaces B: Biointerfaces 107, 257–261.

Yang, T., Wang, S., Jin, H., Bao, W., Huang, S., and Wang, J. (2013c). An electrochemical impedance sensor for the label-free ultrasensitive detection of interleukin-6 antigen. Sensors and Actuators B: Chemical 178, 310–315.

Yu, X., Xu, D., Lv, R., and Liu, Z. (2006). An impedance biosensor array for label-free detection of multiple antigen-antibody reactions. Front Biosci 11, 983–990.

Yun, C., Ya, Y., and Yifeng, T. An electrochemical impedimetric immunosensor for ultrasensitive determination of ketamine hydrochloride. Sensors and Actuators B: Chemical.

Yun, Y., Bange, A., Heineman, W.R., Halsall, H.B., Shanov, V.N., Dong, Z., Pixley, S., Behbehani, M., Jazieh, A., Tu, Y. et al. (2007). A nanotube array immunosensor for direct electrochemical detection of antigen–antibody binding. Sensors and Actuators B: Chemical 123, 177–182.

Zelada-Guillen, G.A., Tweed-Kent, A., Niemann, M., Goeringer, H.U., Riu, J., and Xavier Rius, F. (2013). Ultrasensitive and real-time detection of proteins in blood using a potentiometric carbon-nanotube aptasensor. Biosens Bioelectron 41, 366–371.

Zhang, H., Wang, R., Tan, H., Nie, L., and Yao, S. (1998). Bovine serum albumin as a means to immobilize DNA on a silver-plated bulk acoustic wave DNA biosensor. Talanta 46, 171–178.

Zhang, K., Ma, H., Zhang, L., and Zhang, Y. (2008). Fabrication of a sensitive impedance biosensor of DNA hybridization based on gold nanoparticles modified gold electrode. Electroanalysis 20, 2127–2133.

Zhang, X.E. (2000). Screen-printing methods for biosensor production. In Biosensors, J. Cooper, and A. Cass, eds. (Oxford: Oxford University Press).

Zheng, H., and Du, X. (2013). Reduced steric hindrance and optimized spatial arrangement of carbohydrate ligands in imprinted monolayers for enhanced protein binding. Biochimica et Biophysica Acta (BBA) - Biomembranes *1828*, 792–800.

Ziegler, C. (2004). Cantilever-based biosensors. Anal Bioanal Chem *379*, 946–959.

Zourob, M., Elwary, S., and Turner, A.P.F. (2008). Principles of Bacterial Detection: Biosensors, Recognition Receptors, and Microsystems (Springer Science+Business Media, LLC).

www.ingramcontent.com/pod-product-compliance
Lightning Source LLC
Chambersburg PA
CBHW050452190326
41458CB00005B/1246